The Bloomsbury Series in Clinical Science

Titles in the series already published:

Bronchoalveolar Mast Cells and Asthma
K. C. Flint

Platelet-Vessel Wall Interactions
Edited by R. Michael Pittilo and Samuel J. Machin

Oxalate Metabolism in Relation to Urinary Stone
Edited by G. Alan Rose

Diseases in the Homosexual Male
Edited by Michael W. Adler

Herpes Simplex Virus
Adrian Mindel

Parasitic Disease in Clinical Practice
G. C. Cook

Lasers in Urology: Principles and Practice
Edited by T. A. McNicholas

Forthcoming titles in the series:

The Blood Brain Barrier for Clinicians
Alan Crockard and Nicholas Todd

The Parotid Gland
M. Hobsley and G. T. Watkin

Medical Problems in AIDS
Edited by Ann Millar

HEARING LOSS IN THE ELDERLY

AUDIOMETRIC, ELECTROPHYSIOLOGICAL AND HISTOPATHOLOGICAL ASPECTS

Sava Soucek and Leslie Michaels

With 52 Figures

Springer-Verlag
London Berlin Heidelberg New York
Paris Tokyo Hong Kong

Sava Soucek, MD, PhD
Consultant in Audiological Medicine, Ear, Nose and Throat
Department, St Mary's and Central Middlesex Hospitals, London; and
Honorary Senior Lecturer in Audiological Medicine, St Mary's Hospital
Medical School (University of London), UK

Leslie Michaels, MD, FRCPath, FRCP(C), FCAP, D.Path
Professor of Pathology, Department of Histopathology, University
College and Middlesex School of Medicine (University of London),
Institute of Laryngology and Otology; and Honorary Consultant
Pathologist, Royal National Throat, Nose and Ear Hospital and
Bloomsbury Health Authority, London, UK

Series Editor
Jack Tinker, BSc, FRCS, FRCP, DIC
Postgraduate Medical Dean, British Postgraduate Medical Federation,
University of London, 33 Millman Street, London WC1 3EJ, UK

ISBN-13:978-1-4471-1807-7   e-ISBN-13:978-1-4471-1805-3
DOI: 10.1007/978-1-4471-1805-3

Cover: Fig. 4.7 (giant stereociliary degeneration); the graph is adapted
from Fig. 3.20 (cochlear microphonics).

British Library Cataloguing in Publication Data
Soucek, Sava
Hearing loss in the elderly: audiometric,
electrophysiological and histopathological aspects.
1. Old persons. Hearing disorders
I. Title II. Michaels, L. (Leslie) *1925–* III. Series
618.9778
ISBN-13:978-1-4471-1807-7 W. Germany

Library of Congress Cataloging-in-Publication Data
Soucek, Sava, 1935–
Hearing loss in the elderly: audiometric, electrophysiological
and histopathological aspects/Sava Soucek and Leslie Michaels.
p. cm. Includes bibliographical references
ISBN-13:978-1-4471-1807-7
1. Presbycusis. I. Michaels, L. (Leslie) II. Title. [DNLM:
1. Deafness—in old age. 2. Hearing Disorders—in old age.
WV270 S719h] RF291.5.A35S68 1990 618.97'78—dc20
DNLM/DLC
for Library of Congress   90-9929
                                                                 CIP

Typeset by Best-Set, Hong Kong

2128/3916–543210   Printed on acid-free paper

Dedicated to the memory of the late Norman Exton-Smith, Emeritus Barlow Professor of Geriatric Medicine, University of London, who fostered this work throughout its whole course

Series Editor's Foreword

In their preface, the authors highlight the great social and medical importance of hearing loss in the elderly, an ever-increasing problem. It is, therefore, most appropriate that a monograph, devoted to research in this field, features as an addition to the Bloomsbury Series in Clinical Science. Written by two leading authorities, the book reviews the past and present literature, details the clinical aspects and considers the electrophysiological and histopathological issues basic to the overall problem. It continues the high standard and excellence fostered by previous issues in the Series.

London, August 1990   Jack Tinker

Preface

The number of very old people in the community is steadily increasing so the hearing loss that many of them endure has become a disorder of social and medical importance. Only by a greater scientific knowledge of old age deafness can there be any possibility for improvement in the care offered to its sufferers.

The work presented here is a clinicopathological research report in this field in which we have attempted to enhance understanding by addressing two main questions:

1. Where does the disturbance giving rise to old age deafness reside?
2. What is its cause?

It is now generally accepted that the pathological changes giving rise to deafness in old age lie in the cochlea, but at the same time the tenet has become embedded in medical literature, derived from histological interpretations in one centre, that there are four different sites of the lesion in the cochlea, giving rise to four different forms of presbyacusis. We have studied the functional aspects of this problem and also, by analysing adequately prepared material only, subjected the morphology to our own independent scrutiny. Many workers still believe that presbyacusis may have its cause in different systemic diseases arising outside the ear. Our work exposes this idea also to critical analysis.

Most of the scientific work on presbyacusis is now carried out as observations on ageing animals. While the findings of such work are of interest they do not necessarily contribute to the elucidation of human presbyacusis. We have chosen in the studies recorded here to confine ourselves to the human disease.

We were able to analyse successfully the electrophysiological processes in the human auditory system with the skilled advice of Steve Mason PhD, Principal Physicist at the Queen's Medical Centre, University of Nottingham and his presence as a co-author of the chapter on this subject acknowledges this close collaboration.

Acknowledgements

We wish to thank Tony Frohlich for his valuable help in preparation of temporal bone specimens and Andrew Gardner and Caroline Lonnen for the artwork and photography.

The studies were supported by a grant from the Royal National Institute for the Deaf. Duphar Laboratories, Southampton, UK, generously covered the cost of the publication of colour prints.

London   Sava Soucek
April 1990   Leslie Michaels

Contents

Contents xiii

Review of Literature

Up to the Early Twentieth Century

The scientific recording of deafness concomitant with old age commenced only in the past century-and-a-half because the numbers of sufficiently old individuals available for observation may not have been substantial until recent times. We have not been able to find a record of an examination of the functional and morphological basis of this deafness until that of Toynbee in 1849. The clinical tests used in this study were crude, the mainstay for the diagnosis of deafness being the sensation of the ticking of a watch held at 4 feet from the subject. No observations were made on changes in the hearing of sounds at different pitch. Only gross, but no microscopical, observations were made in the morphological part of the study, which was a dissection of the temporal bone in 18 cases. The wrong part was studied, the middle ear, not the inner ear, and no adequate control observations were made. Hence the finding of an increased thickness of the mucosa of the middle ear including that of the tympanic membrane in the elderly patients, was worthless as an explanation of old age deafness. Even more open to criticism was the author's attempt to alleviate the deafness, presumably in the hope of reducing the thickness of the affected parts, by pouring solutions of silver nitrate or mercurous chloride into the ear canals of his hapless patients!

Some 41 years later, an understanding of the correct basis of old age deafness began to emerge. This was the result of the studies of Zwaardemaker in Utrecht (1891). He was an otologist who routinely used in his clinic devices known as "Galton's whistles" to test his patients' hearing. These were tin whistles of various lengths (Fig. 1.1). The length of each whistle gave an approximate measure of its relative pitch, and the upper limit of frequency that the subject could hear was recorded as the length of the shortest whistle within his range of hearing. By testing many subjects Zwaardemaker was able to construct a scale of progressively lengthening whistles which constituted the upper limits of hearing for different ages, from childhood to extreme old age.

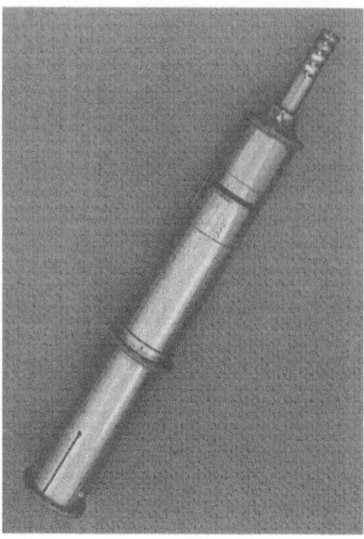

Fig. 1.1. Galton's whistle. The frequency of the tone emitted by blowing at the end of the whistle is dependent on the length. Zwaardemaker employed a series of whistles of different lengths to test patients with old age deafness.

In a later publication (Zwaardemaker 1894) he indicated that the shortening of tone in old age was equivalent to the whole of the uppermost musical octave, so that children have a total range of hearing of 11 octaves while elderly people have only 10 octaves. He was able to exclude middle ear disease as a basis for the high tone hearing loss of the elderly, because he observed that patients with such middle ear disease did not have such a loss. He felt that the origin of the old age disorder must lie in the "labyrinth", and in support of this he noticed also that conduction of sound through the skull bone was diminished in the elderly. Credit must, therefore, be given to Zwaardemaker for these fundamental observations which cleared the way to later productive research on the sensorineural loss of hearing in old age.

Nineteenth century medical writing is otherwise silent on the deafness of old age and the textbooks exhibited a characteristically long delay before even Zwaardemaker's discovery reached them. Thus the monographs on ear disease of Wilde (1853) and Gruber (1893) and all the editions of Politzer's textbook up to the last in 1926 are free of any discussion on the subject. The first textbook in which old age deafness is mentioned is that of McKenzie (1927). Why were the basic facts relating to presbyacusis not revealed until the end of the nineteenth century?

The nature of sound and the basis of pitch in the frequencies of its vibrations was known to early nineteenth century science. The tuning fork as a means of producing tones at differing vibrations was in fact recommended in the 1840s and its use in testing air and bone conduction was described as a method of discriminating sensorineural deafness at that time (Despretz 1845). The delay between the scientific discoveries on the one hand and the use of quantitative and other techniques based on them in clinical research and practice on the other is well demonstrated in the case of old age hearing loss. In the interval between Toynbee's and Zwaardemaker's studies scientific understanding of tone perception had been advanced by von Helmholz (1863) and the detailed histology of the end organ of tone reception was worked out by Retzius (1881–1884). The physiological basis of sound reception having become so well established, it was natural for a practising otologist to take up the concept of frequency estimation and to use it as Zwaardemaker did to test the hearing of his patients at all ages. The reason for the lag in understanding of the real nature of presbyacusis was thus the pause between laboratory advances in physiological acoustics and the clinical research that was necessary to make the discovery. Progress in basic science does not automatically lead to clinical advances. These can frequently only be achieved by a clinical observer familiar with the basic science who asks the right questions – a lesson sometimes forgotten by modern research grant-providing agencies.

Further advances in acoustical science and technology led from Zwaardemaker's testing of pitch perception by simple whistles of specific lengths to the production of pure tones by electrical means and the possibility of delivery of accurately measured variations in the intensity of the different tone levels. So Zwaardemaker's discovery of the "law" of diminution of high tone hearing with advancing age could now be confirmed and refined by the use of electrical audiometers producing modern type audiograms (Bunch 1929).

Recent Times

After recognition of old age deafness as a sensorineural hearing loss particularly for high tones, interest in this condition increased and has progressed in logarithmic proportions until the present time. It is likely that this activity has been stimulated by the growth in numbers of old people in the community.

The literature contains many imputations about the disease process in both its functional and in its morphological aspects. Conclusions reached are frequently contradictory from one paper to another. Faced with such a daunting volume of writings on the subject it was necessary to define a programme by which to survey it. In the following account the pathways in the literature on the subject are mapped out under several headings both to provide an intelligible review of the literature as a whole and a backcloth for the original observations described in the subsequent sections of this monograph.

A large part of the literature has been concerned with studies to define the site of the pathological change which is at the basis of old age hearing loss. Broadly speaking such studies may be divided into two groups:

1. Those that are concerned with the possibility of cerebral (including brainstem) disturbances as the basis of old age hearing loss. These studies have been mainly of functional aspects in living people. A few have been of a morphological character and carried out on the brain at post-mortem.

2. Those that are concerned with abnormalities of the peripheral part of the auditory system as a basis for presbyacusis. In contrast to the central studies those in this group are predominantly morphological in nature, involving post-mortem analyses of cochlear pathology in most cases.

The purpose of many investigations has been to attempt to discover an aetiological basis for old age deafness. Two contrasting tendencies allow these works to be divided into two groups:

1. Those that indicate that old age deafness may arise from noxious features of the environment affecting the auditory system, either outside the body, such as noise, or within the body such as cardiovascular disease or hyperlipoproteinaemia. In the view of these authors presbyacusis is exogenous in origin.

2. Those that indicate that old age deafness arises as an innate disturbance of part of the auditory system, occurring as a degeneration of old age, unaffected by outside influences. These authors regard presbyacusis as endogenous in origin.

Localization of the Disturbance

Central

Functional Aspects

Many tests have been devised which purport to distinguish between auditory damage of peripheral (cochlear) origin on the one hand from central (cerebral) origin on the other. The essence of the performance of such tests is the delivery of messages which require central processing in their deciphering. It has been claimed in a number of studies that lack of central auditory ability plays an important part in the pattern of old age deafness. In the study of Welsh et al. (1985), for instance, 72 elderly individuals in three groups of mean ages 65.3, 75.2 and 83.8 years were subjected to a battery of tests of central auditory control, including compressed speech, competing sentences, low pass filtered speech, binaural fusion and rapidly altering speech perception. It was found that there was an increasing failure rate in these tests with ageing. Another means of assessing central disturbance in presbyacusis has been that involving the use of the reaction time, which has been assumed to be centrally rather than peripherally determined. Maspetiol and Semette (1968) and Feldman and Reger (1967) have used three aspects of the reaction time for study: visual, tactile and auditory. In each stimulus modality the reaction time increased with advancing age. For visual and tactile reaction time the increase from decade to decade was relatively linear whereas auditory reaction time appeared to increase disproportionately compared to the other modalities. It has even been suggested that the degenerative changes which occur in the brain in ageing are the lesions primarily responsible for the overall picture of presbyacusis (Hinchcliffe 1962). It must be realized, however, that these central tests analyse not the auditory functions but only the integrative and reactive functions of elderly patients. These may well be disturbed in subjects who have concomitant but unrelated hearing problems; the tests may, therefore, be positive in elderly patients, but would not necessarily throw any light on the location of the essential lesion of presbyacusis.

Indeed earlier work carried out by one of us (S.S.) and reported in Czech has indicated that central disturbances play positively no part in presbyacusis.

The work involved a quantitation of the usual techniques of speech audiometry (Součková 1973). The speech material offered to the subjects in these tests was selected in such a way that it was possible to distinguish a lesion in the peripheral part of the auditory system, where information is coded, from one in the central part where semantic information is integrated. Using this method elderly people aged 61 to 80 years (average age 79.2 years) in residential homes were investigated. They were examined otoscopically and by pure tone audiometry followed by speech audiometry monaurally. The findings in this study suggested that the perception of speech appeared to be damaged from a deficiency in coding of phonemes in the peripheral part of the hearing analyser and not from any central deficit.

Our experience was then biased towards a peripheral location as the seat of old age hearing loss when we commenced the work to be described in subsequent sections and we gave no consideration to the possibility of using speech audiometry or other forms of central testing as part of this work. This attitude has been strengthened by recent findings suggesting that phonemic regression is not a concomitant of ageing (Holmes et al. 1988). Audiometry of any type also demands close co-operation of the subject, which is not easy to attain in elderly people. This aspect will be discussed below.

Morphological Aspects

In Brain. There have been few studies of changes in the cerebral component of the auditory system in the aged. The cochlear nuclei have been most studied because of the easy access and fixation of these structures which lie superficially in the medulla. Arnesen (1982) counted the nerve cells in the cochlear nuclei in six cases aged 76 to 89 years in whom audiograms had indicated old age deafness. He did not carry out his own control counts on normal brains, but compared his findings with the nerve cell numbers obtained by Hall (1964) from the cochlear nuclei of asphyxiated neonates. Although Arnesen took care to make his own technique as close as possible to that of Hall, the lack of direct control cases prepared and examined in the same laboratory must put his findings of a loss of 50% of nerve cells into question. Konigsmark and Murphy (1972) did examine their own control material in a study of the cerebral cochlear nuclei in subjects from the newborn to 90 years of age, but found no reduction in nerve cells in the aged specimens. The latter, unlike Arnesen's cases, had not been investigated by audiometry and it was not known whether any of them had presbyacusis. In view of our investigations, in which deafness was found to be universal in elderly subjects (see Section 2), it would seem that the findings of Konigsmark and Murphy do have importance in showing with adequate controls that old age deafness is not related to degeneration of central cochlear nuclei.

Dublin (1976, 1986) claims to have found evidence of tonotopic representation in the spheroid nerve cells of the human superior ventral cochlear nucleus, the loss of such cells corresponding to the pattern of hearing loss in the audiogram. Since the cochlear nucleus of only a single patient is described in each of his two papers it is difficult to accept the author's claim that histological examination of the cochlear nucleus by this method is useful in the postmortem evaluation of presbyacusis. Hansen and Reske-Nielsen (1965), studying the nuclei qualitatively only, found degenerative changes in the nerve cells of both the ventral and dorsal auditory nuclei in 12 elderly patients most of whom had had typical audiometric features of presbyacusis. It is well known that such degenerative changes are a frequent feature of many parts of the aged brain and often may be the result of the patient's terminal illness. The appearance of degeneration at one particular site in aged patients with presbyacusis cannot be considered as evidence that such changes at that site are the basis of the presbyacusis. In the same work Hansen and Reske-Nielsen describe similar degenerative changes in the cells of the inferior colliculi, medial geniculate ganglia and temporal lobes and also mention arteriosclerosis and arteriolosclerosis of vessels in those areas.

These pathological changes in the higher levels of the auditory system, while of some interest, must be regarded with caution from the point of view of the location of the basic lesions of old age deafness. The few investigations which have been carried out show by no means comparable findings, the numbers of cases investigated being small and the quantitative investigations often not controlled. It seems reasonable to accept from available evidence that there may be some loss of cochlear nucleus cells in some elderly auditory systems, but not in all. The hearing loss is more probably produced by another more consistently and specifically related disturbance, any nerve cell depletion in the central nervous system being a late result of the latter. More centrally than the cochlear nuclei the auditory system presents an extensive and diffuse collection of nerve cells and their processes which are difficult to study and the changes of which are of obscure significance. From the small amount of neuropathological data available it seems unlikely that any central pathological changes will prove significant as a basis for presbyacusis.

Peripheral

The concept of a peripheral, i.e. aural, disturbance as the basis of the hearing loss suffered by the elderly has presented a more realistic possibility to many students of old age deafness. One possible site for the pathological change leading to this disorder in the peripheral part of the hearing system is the middle ear.

Middle Ear

The earliest detailed examination of old age hearing changes, which was mentioned above, was that of Toynbee (1849). In this study the author concluded, on the basis of dissections of 18 middle ears from elderly people, that the affection was one of the middle ear since he found a thickening of the middle ear mucosa and of the tympanic membrane together with bands of adhesions in the middle ear cavity. The study was not controlled by similar observations of patients at a younger age group without hearing loss. Moreover, Toynbee could not have realized that he was looking for the lesions of old age hearing loss at the wrong site. The functions of the different parts of the ear were not known and differentiation between deafness due to middle ear disease and that due to cochlear and retrocochlear disease was not yet possible.

Modern audiological methods have in fact indicated that the middle ear component of old age deafness can be only very slight. Thus Nixon et al. (1962) found air conduction to be reduced in the elderly by a maximum of only 12 dB at 4 kHz compared with bone conduction. The cause of such slight functional changes probably lies in the fibrous and bony ankyloses which have been demonstrated to occur with advancing age in the ossicular articulations (Belal and Stewart 1974). It thus seems definite that middle ear changes cannot play a significant part in old age hearing loss.

Inner Ear

Functional Aspects. Audiometry has been the usual means of studying the functional aspects of hearing loss in the aged. A high tone hearing loss has been the most frequent finding, but some workers have indicated a loss in both high and low tones. Schuknecht (1964) has described three shapes of curve and a pattern of speech audiometry which he has related to different cochlear lesions at post-mortem.

Apart from work done by Schuknecht and others in relating audiometric losses to histopathological changes in the cochlea, investigations using subjective audiometric methods have been carried out along two lines in the study of presbyacusis:

1. Detailed analyses of special audiometric tests to attempt to throw some light on the character and site of the disturbance.

2. Epidemiological surveys in which pure tone audiometry has been used as a means of identifying the incidence, degree and frequency characteristics of hearing losses in a population. A review of this subject will be presented in Section 2.

Few results of special audiometric analyses in old age hearing loss are available. The findings of Lehnhardt (1984) are representative. Recruitment is a subjective phenomenon in which an ear with sensorineural hearing loss seems to hear tones better than the normal ear. A method has been devised – the short increment sensitivity index (SISI) – in which a short increment of intensity imposed on a carrier tone is used as an indicator of recruitment (Beagley and Barnard 1982). A carrier tone starting at 20 dB is given 20 increments of 1 dB, each lasting for a fraction of a second. Normal ears will hear only about 20% of the increments. Lehnhardt states that subjects with presbyacusis usually have almost 100% detection of the increments, indicating a high degree of recruitment and therefore of cochlear damage. on the other hand Lehnhardt finds no tone decay in presbyacusis. This test is conducted by presenting a tone at 10 dB above the threshold for one minute, following which the threshold is tested again to find out if it has worsened. If positive (i.e. the perception of sound has not diminished), this test denotes a retrocochlear hearing loss. That it is negative in presbyacusis (i.e. perception of sound diminishes as in normal ears) would again indicate the intracochlear basis for the disturbance. In the Békésy test Lehnhardt finds the separation between pulsed and continuous traces to be no more than 20 dB, which would correspond to the type II variety of tracing (Beagley and Barnard 1982). This again is indicative of a cochlear source of the hearing loss.

It must be pointed out that audiometric tests require the close co-operation of the subject. The special tests just described demand a considerable degree of concentration. Old people frequently lack concentration and co-operate in tests with difficulty, the degree of this disability increasing with the age of the patient. Special audiometric tests carried out on younger sufferers with presbyacusis who do not show the full severity of the disease, do not indicate the outstanding features of the full-blown condition. On the other hand, those with marked features of the disease are often too old to participate in tests of this type.

Electrophysiological tests, such as brainstem evoked responses and electrocochleography do not require the positive co-operation and concentration of the subject and so are more suitable functional investigations for the study of old age hearing loss in its severe form. The results of studies which we carried out using such electrophysiological techniques will be described in Section 3

Morphological Aspects. Many studies have been concerned with the morphological changes in the cochlea that might constitute the pathological basis for hearing loss in the aged. There has been a surprising lack of agreement among them as to the most important change. For this reason the subject will be considered under the following specific anatomical designations, and those studies which consider the particular site or sites as being the important source of the disturbance in presbyacusis will be discussed:

(a) Hair cells

(b) Spiral ganglion cells and nerves

(c) Stria vascularis

(d) Multiple cochlear sites

(e) Bone in fundus of the internal auditory meatus

(a) *Hair cells.* Degeneration of the hair cells has been blamed by some for the hearing defect occurring with increasing age. However, only the studies of Bredberg (1965) and of Johnsson and Hawkins (1967) give sufficient evidence to incriminate these structures as the primary site of the defect. Both of these investigations used surface preparations of the human cochlea to identify the loss.

Bredberg (1965) examined the hair cells of the cochlea in 29 fetuses, six infants and 72 subjects of ages ranging between 1 and 93 years. Fixation was performed by perfusion of the perilymphatic space with fixative via the round window soon after death. The temporal bone was then removed from the skull and the bony labyrinth expunged by drilling. The organ of Corti was then examined microscopically in phase contrast light after staining with osmic acid and placing portions of the basilar membrane flat onto slides.

A clear reduction of hair cells with advancing age was found by Bredberg. The inner hair cells were lost mainly in the basal region of the cochlea, but the outer hair cells were lost all through the organ of Corti, with a particularly marked disappearance of cells at the base and at the apex. Nerve fibres in the osseous spiral lamina of the basal coil were usually reduced or absent in old age, and a similar change was sometimes present in a patchy form in other parts of the cochlea in association with loss of inner and outer hair cells. Sixteen of Bredberg's subjects had had pure tone audiometry and he related the high tone deafness to the basal coil hair cell losses and nerve fibre disappearance which were displayed in the corresponding cochlear surface preparations.

We believe that Bredberg's is the crucial study in the definition of the morphological basis of old age hearing loss. It takes post-mortem cochlear autolysis into account, considers the difficulties of inspecting hair cells in histological section and uses methods which avoid them. It clearly relates the changes detected to the functional alterations. It indicates a lesion which is prevalent throughout the older subjects and which is progressive from childhood to old age. The observations also provide an explanation for the basal coil spiral ganglion loss which has been found in many, *but not all*, elderly cochleas (see below).

Johnsson and Hawkins (1967) made a similar study, surveying 150 subjects from fetuses to 97 years of age. Counts of hair cells were not made. Observation of a definite progression of hair cell and nerve degeneration from the extreme end of the basal turn to involve a major part of the basal turn with advancing age was made and the relation of nerve degeneration to preceding hair cell loss was clearly established. No audiograms were available.

In other studies hair cell loss has been described on the basis of observations of histological sections (Schuknecht 1955, 1974; Suga and Lindsay 1976). Statements about the appearances of hair cells in histological section must be regarded with caution since these cells are very liable to undergo post-mortem autolysis (Eckert-Mobius 1926). Moreover, representation of the functionally most important part of the hair cells, the stereocilia, in a histological section is very small even when adequately fixed.

(b) *Spiral Ganglion Cells and Nerves.* Clear-cut losses in the spiral ganglion cells and the nerve fibres derived from them are well-documented in the literature on old age hearing loss (Guild et al. 1931; von Fieandt and Saxen 1937; Fleischer 1956; Hansen and Reske-Nielsen 1965; Suga and Lindsay 1976). The loss when present is always most marked in the terminal part of the basal coil (see Section 4). It is frequently accompanied by nerve fibre degeneration extending along the basilar membrane to the organ of Corti. In some cases a loss of corresponding hair cells has been demonstrated (Saxen 1952), but, since this has been observed in histological section, it is open to the criticism given above and the claim should be treated with some reservation. Saxen (1952) observed an occasional cell showing neuronophagia in the spiral ganglia of elderly cochleas, but other signs of cell degeneration were not present.

The frequency of the observation of spiral ganglion cell loss in presbyacusis gives it a valid claim for consideration as the structural basis for many cases of that condition. Unlike hair cells, spiral ganglion cells are well seen in histological sections of the human cochlea and do not suffer the same tendency to autolytic change. We would suggest that it is for precisely these reasons that spiral ganglion cells have received most attention. Spiral ganglion cell loss is not found in all elderly cochleas, however. It seems more likely that hair cell loss, found by Bredberg (1965) using the technically superior method of surface preparation, to be present in *all* elderly cochleas, is the primary lesion of presbyacusis. This lesion has, we feel, been overlooked in some cases because the hair cells had been examined by histological sectioning only.

(c) *Stria Vascularis.*   Atrophy of the stria vascularis has been stated to be the pathological basis of old age deafness in some subjects (Schuknecht 1964). Like the hair cells the stria is subject to post-mortem change as a result of which the appearance may resemble atrophy. Vascular thickening in the stria has been incriminated by some, but its importance has been difficult to assess (von Fieandt and Saxen 1937; Jørgensen 1961).

(d) *Multiple Cochlear Sites.*   A concept has gained ground, and is now favoured as the main one defining the pattern of old age hearing loss, that there are separate types corresponding to specific pathologies at each of the above sites, i.e. hair cells (sensory presbyacusis), spiral ganglion cells (neural presbyacusis), stria vascularis (strial or metabolic presbyacusis) and a further type "cochlear conductive presbyacusis" which is thought to be due to stiffening of the basilar membrane (Schuknecht 1974). Each type of presbyacusis is considered to be associated with its own audiometric feature: sensory presbyacusis with an abrupt high tone loss in the audiogram, neural presbyacusis with a loss of speech discrimination, metabolic presbyacusis with a flat audiometric pattern and cochlear conductive presbyacusis with a straight line descending audiometric pattern.

The concept is an attractive one in allowing the described pathological features of old age hearing loss to be brought into a single classification. We would suggest, nevertheless, that the multiple type hypothesis needs to be tested and retested before it sinks into the realms of undisputed dogma (as it already shows evidence of doing) for the following reasons:

1. The numbers of cases with a complete history and temporal bone pathology are too small for the hypothesis to be accepted as proved.

2. Vast numbers of pure tone and speech audiograms are carried out in clinical investigation of elderly subjects, but support has not come from the ranks of audiologists for any clear-cut division of cases on the audiometric basis laid out in this classification.

3. The pathological basis of three of the four suggested types of presbyacusis is open to doubt. As mentioned above, the identification of hair cell loss (the feature of sensory presbyacusis) is doubtful in histological sections largely because of the frequent presence of autolysis. The identification of strial atrophy (metabolic type) suffers from the same difficulty. Cochlear conductive damage is still admitted to be a theoretical concept by its proponents.

(e) *Bone in Fundus of the Infernal Auditory Meatus.*   Presbyacusis has been ascribed to hyperostosis of the fundus of the internal auditory meatus (Šercer and Krmpotič 1958; Krmpotič-Nemanič et al. 1972). These changes were stated to progress slowly with age resulting in a reduction of the number of foramina for nerve fibres in the spiral tract. They are said to occur first in the region of the basal coil, thus accounting for loss of high tones at an advanced age. Nerve damage is not described by the authors at the sites of the purported compression. In spite of the 30 years that have elapsed since it was first presented the concept has not yet found favour as an explanation for presbyacusis.

Aetiological Basis for the Disturbance

Role of Exogenous Factors

Hearing loss in the elderly although it is admitted to be a common disorder is currently believed by many to be the result of factors outside the auditory system. Some of these are thought to be derived from outside the body altogether, such as noise or diet; others are pathological changes in other systems which are considered to act on the auditory system secondarily, such as diabetes or cardiovascular disease. Here the literature describing the role of the two groups of possible exogenous conditions will be considered critically.

Noise

It is certain that severe and permanent hearing loss can be produced by exposure to noise. A proportion of deaf elderly people have a hearing loss which has been caused by a noisy occupation. Existence in the twentieth century has provided many opportunities for noise exposure apart from occupational ones. The blare of radio, television and gramophone, the noise of traffic, the drilling of road menders – each hour provides a different source of severe noise which impinges on most people in their non-working environment. It has been suggested that these daily insults to the hearing can mount up throughout a lifelong exposure and cause sufficient damage to lead to an extensive degree of hearing loss in old age, which then becomes manifest as presbyacusis.

It must be said that the changes in the inner ear following high intensity sound waves appear similar to those described by Bredberg (1965) as being characteristic of old age hearing loss (see above). In noise-induced hearing loss the alterations involve mainly the outer hair cells of the basal coil. Inner hair cells and higher cochlear coils may also be affected following greater degrees of noise (Paparella and Melnick 1967). It is possible that the similarity between the cochlear changes after excessive noise and those reported by Bredberg in old age is due to the fact that the organ of Corti can only respond to different damaging processes in a limited fashion. It is also possible that noise is an underlying cause of old age deafness.

Several attempts have been made to answer the question as to whether the noise of everyday life is a cause of presbyacusis by testing the hearing of populations living throughout their lives in remote areas away from the pandemonium of modern civilization. An audiometric examination of 541 unselected subjects from 10 to 90 years of age living in a remote area of the Sudan about 650 miles southeast of Khartoum was carried out by Rosen et al. (1962). The population lived in bush country only accessible by truck or jeep during the dry season. These people, the Mabaans, are exposed only to rural sounds, the loudest of which is that of the lowing of cattle or occasional claps of thunder. Hearing tests were carried out with battery operated audiometers. The number of subjects tested over the age of 60 and up to 80 years was 124. The audiometric findings were compared with those of a sample of the US population measured by Glorig and Nixon (1960). The Mabaans seemed to demonstrate better hearing in the higher frequencies than those of that report or of any similar studies emanating from the inhabitants of modern Western civilization.

An important consideration which makes it difficult to accept the results of this study is the methods used to determine the ages of the Mabaans tested. Birth certificates were, of course, not available. The ages of the children were determined by the investigators from signs of puberty and eruption of teeth, but in the elderly group under consideration there was no such physiological assessment available and only indirect markers of age such as the ages of children and grandchildren of the subjects were available. Under such circumstances the relationship of hearing loss to age must be difficult to assess and any comparisons of such losses with populations of certificated ages should be considered circumspectly.

It was further pointed out by Bergman (1966), who was one of the team that tested the aboriginal population, that the average Mabaan in the older decades hears no better than the top 10% of the US population. Bergman suggested that there might be a homogeneity of the Mabaan population with regard to old age deafness as compared with a heterogeneity of the American one. The Mabaan might have a genetically lesser tendency for hearing to deteriorate in old age. Americans, too, might comprise some people of this type, but also many with a greater tendency for hearing loss in old age. Old age hearing loss would then not necessarily be the result of the presence of noise in the environment (extrinsic), but of constitutional make-up (intrinsic).

In a study of the hearing thresholds of an island population in the northern part of Scotland, Kell et al. (1970) found that the hearing acuity was even better than that of the Mabaans until the age of 60 was reached, after which it was inferior. The hearing levels were superior to those of the British Standard for normal hearing, but similar to those of two other Scottish studies. Again it is possible to explain differences such as these on a constitutional basis rather than on noise exposure. The Scottish were certificated for age and thus difficult to compare realistically with the Mabaan group. Observations of the hearing of elderly Easter islanders (natives of a remote and quiet Pacific island) showed that there was some old age hearing loss (Gooycoolea et al. 1986). Only 23 native people were involved in this study, however, and these were not compared with people who had lived in the much more noisy continental Chile.

In our opinion environmental noise has not yet been proved to be a factor in old age hearing loss. Too much credence has been given to the apparent difference between people living in remote noise-free environments and those in urban settings. The methods used by the more primitive arms of such investigations are not yet sufficiently sophisticated to yield statistically valid conclusions. It has become a danger with the spread of civilization that people living in sufficiently remote areas may not longer be available for such studies even if objective methods of assessing their ages could be devised.

Cardiovascular Disease

A strong impression has developed, with no scientific basis, that cardiovascular disease is a cause of hearing loss in old age. This was an early concept suggested in autopsy reports (Alexander 1902) and favoured by an occasional later study. A histopathological examination of temporal bones, brain and kidneys of 40 patients over 50 years of age, for instance, related hearing loss, luminal narrowing of the internal auditory artery, spiral ganglion atrophy, angiosclerotic changes of the circle of Willis and encephalomalacia (Makishima, 1978). This apparently mammoth work is not enlightened by any photomicrographs or display of quantitative method.

The expedition of Rosen and colleagues to the region of the Mabaans (Rosen et al. 1962) brought the idea of deafness associated with cardiovascular disease to a clinical level and it was refined in the course of further visits (Rosen et al. 1964; Rosen and Olin 1965). In their expeditions Rosen and his colleagues were impressed not only by the good hearing of the primitive tribe in old age, but also by the lack of hypertension and the "probable" lack of arteriosclerosis in the vascular trees of the natives; they found it natural to connect the good hearing with the apparent lack of cardiovascular disease. Autopsies were not carried out, so the investigators were ignorant as to how much arteriosclerosis was indeed present in the population.

The hypothesis that cardiovascular disease is related to old age hearing loss was put to the test by further observation in Finland on the one hand and Yugoslavia on the other (Rosen and Olin 1965). In Finland, coronary heart disease was found to be very prevalent and so was old age hearing loss. In a mountainous area on the Dalmatian coast of Yugoslavia, coronary heart disease was unusual and the hearing of young people aged 10 to 19 years was better than that of young Finns.

In these and some subsequent reports the evidence for any relationship between cardiovascular disease and old age hearing loss is tenuous. As indicated above, it is difficult to establish significant differences between the degrees of hearing losses in the various communities tested. Even if such difference could be established, any correlation with cardiovascular disease might be fortuitous and based on an "intrinsic" rather than an "extrinsic" association. For example, a population might have an inborn tendency to severe hearing loss in old age and also to a heavy incidence of coronary disease. The association of the two conditions epidemiologically would not mean that the coronary disease caused the hearing loss. Indeed if cardiovascular disease caused hearing loss we should expect that medical wards would be filled with deaf people.

Hyperlipoproteinaemia

High levels of blood lipids have been incriminated as a factor in the production of presbyacusis. The lipids that are most amenable to accurate analysis are cholesterol and triglycerides. These substances do not circulate freely in solution in the blood, but are transported in the form of lipoprotein complexes grouped into the following families: chylomicrons, very-low-density lipoproteins (VLDL), intermediate-density lipoproteins (IDL), low-density lipoproteins (LDL) and high-density lipoproteins (HDL). The current classification of conditions in which one or more of the lipoproteins is elevated in concentration in the plasma is given in Table 1.1.

In the visit of Rosen and his colleagues to the Mabaan tribe in the Sudan (1962) the investigators were impressed by their frugal diet with its lack of meat, and it was suggested that their low mean cholesterol level was related to this. Their apparently superior hearing level, as mentioned above, was said to be associated with a lack of arteriosclerosis as a result of their diet (Rosen and

Table 1.1.   Types of hyperlipoproteinaemia

Type I	Increased chylomicrons
Type IIA	Increased LDL
Type IIB	Increased LDL and VLDL
Type III	Increased IDL
Type IV	Increased VLDL
Type V	Increased VLDL and chylomicrons

For explanation of abbreviations please see text.

Olin 1965). Rosen and Olin also studied the serum cholesterol levels, electro-cardiograms and hearing tests in two mental hospitals in Finland over a period of 5 years. In one hospital the diet was the traditionally Finnish one containing a large amount of saturated fat. In the other the patients were given more polyunsaturated fats in their diet. The blood level of cholesterol dropped in the latter by a statistically significant amount and the hearing thresholds improved considerably throughout the entire audiogram range when compared to the control patients. Patients aged 50 to 59 in the experimental (dieted) hospital had better hearing than those 10 years younger in the control (non-dieted) hospital. There has been no comparable study relating cholesterol levels to hearing loss in a group not complaining of hearing loss, before and after a low cholesterol diet, in spite of the vast amount of research that has been carried out on cholesterol and diet in recent years.

Spencer (1973) in a series of 444 cases with cochlear or vestibular disease found that 46.6% had a significant hyperlipoproteinaemia. Most of these were classified as type IV and smaller numbers as types IIA, IIB and III hyperlipoproteinaemia (Spencer 1975). The symptoms of some patients, including those of the author himself, improved after dieting. A criticism of this study is the lack of normal levels provided with regard to the laboratory concerned and the methods which were adopted, and also the lack of statistics for normal levels of these substances in the same geographical area.

In spite of the lack of firm evidence of the association of hyperlipoproteinaemia with old age hearing loss, support has been given to this concept by many prominent investigators (e.g. Gilad and Glorig 1979). It is well established that the majority of subjects over the age of 70 have a significant hearing loss which is more severe that that of the previous decade (see Section 2). It would be expected that if lipoprotein levels in the plasma were related to presbyacusis the corresponding levels in the general population would, like the hearing loss, steadily increase with age. Examination of the data in the Lipid Research Clinics Population Data Book (1980) show that, although the plasma LDL cholesterol level in 11 North American white populations does increase significantly in each decade from childhood to ages 65–69, the mean level of this substance is significantly reduced above the age of 70 years. A similar result obtains for the plasma triglyceride level, except that elevation occurs until age 54 years and thereafter its level becomes reduced. These findings do not support a relationship between blood lipids and old age deafness. However, in view of the widespread impression that such a relationship does exist, further study of this matter is indicated.

Blood Viscosity

Viscosity is a form of internal friction in a fluid in which the sliding of one part of the fluid over another is resisted. It is measured as the coefficient of viscosity or shear stress in which the velocity of the fluid is quantified in relation to the external force that gives it motion. Plasma viscosity measurements are sometimes used in medicine in cases of macroglobulinaemia in which symptoms may be produced as a result of excessive viscosity of proteins, usually produced by plasma cell neoplasms.

With the premise that hearing loss of unknown origin, including presbyacusis, may be related to ischaemia and yet vascular disease *per se* shows no definite relationship to hearing loss, Browning et al. (1986) recently set out to investigate another aspect of the circulation that might produce ischaemia: the whole blood viscosity. In 49 patients with sensorineural hearing loss, whole blood viscosity (high and low shear) and plasma viscosity were measured. The patients varied in age from 26 to 90 years with an average of 60.5 years. In a further 92 patients varying in age from 20 to 77 years with a mean of 51.9 years the plasma viscosity was measured indirectly after calculating it from concentrations of plasma albumen and globulin.

A negative relationship was found between hearing thresholds and plasma viscosity – the more viscous the plasma, the better the hearing threshold. This applied to both groups of patients. On the other hand, Browning et al. felt, more significantly, the greater the high shear blood viscosity the poorer the sensorineural hearing thresholds. The ratio of high shear blood viscosity (corrected for haematocrit) divided by plasma viscosity is a measure of red cell rigidity or lack of deformability of the red cells under shear. This ratio was found to vary with the hearing threshold at all frequencies to a statistically significant degree. The relationship could be applied only to the first group of patients, since whole blood viscosity measurements were not performed in the second group.

The observations of Browning et al. indicate that further work is necessary to attempt to confirm the relationship, positive and negative, between sensorineural hearing loss and blood viscosity on the one hand, and sensorineural hearing loss and plasma viscosity on the other, respectively. The connection with old age deafness is not clear. Although some of the patients tested were in the old age category the average age of 60.5 years is a borderline level in this respect. If this relationship were confirmed for old age hearing loss after extending the investigation to elderly subjects, it would be necessary to measure specifically the degree of red cell deformability to find out whether this is at fault. Only when these studies have been carried out and a positive result obtained can the question of hyperviscosity of the blood be considered as a serious possible cause of old age hearing loss.

Other Factors

A trend has developed in some centres in which the concept of old age deafness as a purely ageing phenomenon is dismissed in most cases and, instead, other factors such as noise trauma, metabolic derangements, concussive trauma and infections are stressed as the major contributors. When all such conditions have been excluded it is stated that few patients remain in which hearing loss is actually due to the ageing process. Since a considerable part of this monograph will be addressed to the testing of this idea it may be useful at this stage to consider these factors as possible candidates for the causation of the malady.

Among the non-genetic factors stated to be important in the differential diagnosis of presbyacusis, noise has already been considered above. After a head injury a sensorineural type of deafness may develop without any detectable macroscopic damage to the cochlea. Experimental study of anaesthetized animals given a blow to the exposed skull and then allowed to recover, showed a hearing loss of between 3 and 8 kHz and, after a few weeks, a loss of outer hair cells in the upper basal coil region of the cochlea (Schuknecht et al. 1951). Although the occasional case of old age hearing loss may result from such trauma sustained earlier in life it seems unlikely that substantial numbers of elderly subjects will have a hearing loss on this basis.

There are many substances which are known to produce ototoxic injury to the inner ear after being absorbed into the bloodstream. Five classes may be selected from these which are commonly used in clinical practice: (a) aminoglycoside antibiotics, (b) loop diuretics, (c) salicylates, (d) quinine and (e) cytotoxic drugs used in the treatment of malignant disease. Only in the case of aminoglycoside antibiotics do the pathological changes appear similar to those of presbyacusis, i.e. damage to the outer hair cells of the cochlea (Matz and Lerner 1981). It is difficult to accept that in any significant proportion of elderly people with hearing loss the changes could be ascribed to those or indeed to any cytotoxic drugs. Could presbyacusis be due to the damaging effects of some unknown substance on the cochlea? The possibility of saturated fats in the diet being a cause of old age hearing loss has been discussed above on the basis of the production of atherosclerosis with consequent secondary degenerative change in the cochlea. It is feasible that lipids or other substances may act directly in an ototoxic fashion to produce damage to the hearing organ, but direct evidence of this has not been obtained. Lipids are not known to produce this effect under experimental circumstances, nor has any other substance which produces ototoxicity been incriminated in the large numbers of cases which are characteristic of the incidence of old age hearing loss.

Inflammatory damage in the ear usually affects the middle ear. It can be excluded by simple clinical and audiological tests and has been so excluded in large scale studies of old age hearing loss. Occasionally sensorineural deafness is the result of bacterial or viral labyrinthitis, but the incidence of this condition is far too infrequent for it to be incriminated in presbyacusis. None of the pathological changes described in presbyacusis are those of an inflammatory reaction.

Some metabolic and systemic disorders which have been held to be related to presbyacusis are discussed above. For the reasons given it is doubtful whether these or other systemic conditions could be related to old age deafness which so commonly exists in otherwise healthy individuals.

Neoplasms of the ear are unusual. The commonest one to cause sensorineural deafness is acoustic neuroma. This is usually unilateral, unlike presbyacusis, and frequently is associated with vestibular disturbances. Although the lesion should be considered and excluded in every case of sensorineural deafness, neoplasia does not enter into the question of the pathological basis of hearing loss in the elderly.

In cases of deafness in old age based on genetic causes, we are now, for the first time in this section, citing evidence from the literature for the possibility of a group of disorders in which the origin is from within the cells of the auditory system themselves, in this case inherited as a developmental anomaly. In such cases a familial tendency for the condition is noted and the audiogram is said to show a flat or basin-shaped curve with as much involvement of the low frequencies as the high ones (Paparella et al. 1975). The condition is commonly observed in younger people, few reaching old age before its detection. We are unable, therefore, to accept that a genetically based disorder of the auditory system, in the ordinary sense of the term, is responsible for any significant numbers of cases of presbyacusis.

Indeed the assertion of Lowell and Paparella (1977), reiterated by other authors, that "ageing changes *per se* of the inner ear account for little of the observed loss" has not been proved and, in the studies to be described, will be subjected to analysis by actual observations of the disorder. Lowell and Paparella and other proponents of that view are, in our opinion, misinterpreting the clinical findings of this group of patients, who frequently have concomitant illness at several sites. Every subject who is assessed for presbyacusis requires examination of his auditory system and of his general physical condition. Changes found in the auditory system of a specific character may coexist with those of presbyacusis and do not necessarily exclude that condition. Physical changes found outside the auditory system, such as cardiovascular ones, cannot be put forward as the basis of old age deafness without a great deal of further scientific investigation.

A range of genetically based developmental disturbances may exist in humans which is comparable to similar conditions which have been studied in detail in animals. A mutant form of mouse, for instance, shows a normal development of the inner ear for 8–12 days after birth, after which the hair cells of the organ of Corti begin to degenerate and the animals do not hear (Grüneberg 1956). It is conceivable that a human auditory disorder may exist which possesses a similar natural history to that strain of the newborn mouse, except that the onset of the damage to the organ of Corti may occur in later life, even in old age. If found, the basis for the disorder of old age hearing loss would need to be broader than a genetic one, occurring as it does in the majority of people.

Summary

Loss of hearing of high tones was first established by Zwaardemaker (1891) as the basic defect in old age hearing loss.

Localization

Attributions of a central origin for the hearing loss on both functional and morphological grounds may be erroneous. The error could be caused by the presence of central nervous system abnormalities which happen to be concomitant with peripheral defects in many elderly subjects. Speech audiometric studies using special methods to distinguish peripheral from central hearing disorders have not provided evidence for a central disturbance in the elderly hearing function.

Middle ear pathology contributes little to the hearing loss of old people.

Special audiometric methods, particularly those that utilize recruitment, suggest a localization of presbyacutic damage to the cochlea.

Morphological studies of the cochlea indicate that outer hair cell loss is the main lesion and that this is particularly marked in the basal coil. It is not generally accepted that this is the single primary lesion, and spiral ganglion and nerves, stria vascularis, multiple cochlear sites and other peripheral sites are frequently favoured as the sources of old age hearing loss in the modern literature. Much of this literature is based on studies of unperfused, sectioned cochleas which are subject to artefacts and errors of interpretation.

Aetiological Basis

Attempts to incriminate exogenous factors or factors external to the cochlea have not been successful, although many authors cling to such an explanation on grounds which seem to lack a firm scientific basis. Noise, cardiovascular disease, hyperlipoproteinaemia, blood hyperviscosity and even ototoxic drugs are among the explanations brought forward, but without convincing evidence. A pure ageing process obtains little support in the literature, but there is, on the other hand, little evidence against it.

References

Alexander G (1902) Zur pathologischen Histologie des Ohrenlabyrinthes mit besonderer Berück-sichtingung des Cortischenen Organes. Arch Ohrenheilkd 56: 1–23

Arnesen AR (1982) Presbyacusis–loss of neurons in human cochlear nuclei. J Laryngol Otol 96: 503–511

Beagley HA, Barnard S (1982) Manual of audiometric techniques. Oxford University Press, Oxford

Belal A, Stewart TJ (1974) Pathological changes in the middle ear joints. Ann Otol Rhinol Laryngol 83: 159–167

Bergman M (1966) Hearing in the Mabaans. A critical review of related literature. Arch Otolaryngol 84: 411–415

Bredberg G (1965) Cellular pattern and nerve supply of the human organ of Corti. Acta Otolaryngol (Suppl) 236: 1–135

Browning GG, Gatehouse S, Lowe GD (1986) Blood viscosity as a factor in sensorineural hearing impairment. Lancet i:121–123

Bunch CC (1929) Age variations in auditory acuity. Arch Otolaryngol 9: 625–636

Despretz (1845) Observations sur la limite des sons graves et aigus. Comptes rendus hebdomadaire. T.XX. Columella Press, Phoenix Arizona (Quoted in Politzer A (1907) History of otology, vol 1, reprinted and translated)

Dublin WB (1976) The combined correlated audiohistogram. Incorporation of the superior ventral cochlear nucleus. Ann Otol Rhinol Laryngol 85: 813–819

Dublin WB (1986) Central auditory pathology. Otolaryngology-Head Neck Surg 95: 363–424

Eckert-Mobius A (1926) Histologische Technik und normale Histologie II. Post-mortale Veränderungen. In: Henke Lubarsch O (ed) Handbuch der Speziellen Pathologischen Anatomie und Histologie, vol 12. Springer, Berlin, pp. 89–101

Feldman RM, Reger SN (1967) Relations among hearing reaction time and ages. J Speech Hearing Res 10: 479–495

Fleischer K (1956) Histologische und audiometrische Studien über den alternsbedingten Struktur-

und Funktionswandel des Innenohres. Archiv Ohren-usw Heilk u Hals-usw Heilk 170: 142–167

Gilad O, Glorig A (1979) Presbyacusis: the aging ear. Part I. J Am Auditory Soc 4: 207–217

Glorig A, Nixon J (1960) Distribution of hearing loss in various populations. Ann Otol Rhinol Laryngol 69: 497–516

Gooycoolea HG, Rodriguez LG, Farfan GRN, Martinez GM, Vidal R (1986) Effects of life in industrialized societies on hearing in natives of Easter Island. Laryngoscope 96: 1391–1396

Gruber J (1893) A text-book of the diseases of the ear. H.K. Lewis, London

Grüneberg H (1956) Hereditary lesions of the labyrinth in the mouse. Br Med Bull 12: 153–157

Guild SR, Crowe SJ, Bunch CC, Polvogt LM (1931) Correlations of differences in the density of innervation of the organ of Corti with differences in the acuity of hearing, including evidence as to the location in the human cochlea of the receptors for certain tones. Acta Otolaryngol 15: 269–308

Hall JG (1964) The cochlea and the cochlear nuclei in neonatal asphyxia. A histological study. Acta Otolaryngol (Stockh) (Suppl) 194

Hansen CC, Reske-Nielsen E (1965) Pathological studies in presbyacusis. Cochlear and central findings in 12 aged patients. Arch Otolaryngol 82: 115–132

Hinchcliffe R (1962) The anatomical locus of presbycusis. J Speech Hearing Disorders 27: 301–310

Holmes AE, Kricos PH, Kessler RA (1988) A closed-versus open-set measure of speech discrimination in normally hearing young and elderly adults. Br J Audiol 22: 29–33

Johnsson LG, Hawkins JE (1967) A direct approach to cochlear anatomy and pathology in man. Arch Otolaryngol 85: 599–613

Jørgensen MB (1961) Changes of aging in the inner ear. Histological studies. Arch Otolaryngol 74: 164–170

Kell RL, Pearson JCC, Taylor W (1970) Hearing thresholds of an island population in North Scotland. Int Audiol 9: 334–349

Konigsmark BW, Murphy EA (1972) Volume of the ventral cochlear nucleus in man: its relationship to neuronal population and age. J Neuropathol 31: 304–316

Krmpotič-Nemanič J, Nemanič D, Kostovič I (1972) Macroscopical and microscopical changes in the bottom of the internal auditory meatus. Acta Otolaryngol 73: 254–258

Lehnhardt E (1984) Clinical aspects of inner ear deafness. Springer, Berlin (Translated by Langmaid C, Luetgebrune K.)

Lipid Research Clinics Population Data Book (1980) Vol I. The prevalence study. Washington, DC, US Department of Health and Human Services, Public Health Service, NIH Publ No 80–1527

Lowell SH, Paparella MM (1977) Presbycusis: what is it? Laryngoscope 87: 1710–1717

Makishima K (1978) Arteriolar sclerosis as a cause of presbyacusis. Otolaryngology 86: 322–326

Maspetiol R, Semette D (1968) Les testes d'audiometrie tonal dans les atteintes auditives corticales et centrales. Int Andiol 7: 66–76

Matz CJ, Lerner SA (1981) Drug ototoxicity. In: Beagley HA (ed) Audiology and audiological medicine, vol 1. Oxford University Press, Oxford, pp 573–592

McKenzie D (1927) Diseases of the throat, nose and ear. Heinemann, London

Nixon JC, Glorig A, High WS (1962) Changes in air and bone conduction thresholds as a function of age. J Laryngol 76: 288–298

Paparella MM, Melnick W (1967) Stimulation deafness. In: Graham AB (ed) Sensorineural hearing processes and disorders. Little, Brown, London, p 427

Paparella MM, Hanson DG, Rao KN, Ulvestad R (1975) Genetic sensorineural deafness in adults. Ann Otol Rhinol Laryngol 84: 459–472

Politzer's textbook of the diseases of the ear for students and practitioners (1926) 6th edn. Bailliere, Tindall and Cox, London (revised and largely rewritten by Ballin MJ.)

Retzius GM (1881–1884) Das Gehörorgan der Wirbelthiere, vols 1 and 2. Samson and Wallin, Stockholm

Rosen S, Olin P (1965) Hearing loss and coronary heart disease. Arch Otolaryngol 82: 236–243

Rosen S, Bergman M, Plester D, El-Mofty A, Satti MH (1962) Presbyacusis study of a relatively noise-free population in the Sudan. Ann Otol 71: 727–743

Rosen S, Plester D, El-Mofty A, Rosen HV (1964) Relation of hearing loss to cardiovascular disease. Trans Am Acad Ophthal Otol 68: 433–444

Saxen A (1952) Inner ear in presbyacusis. Acta Otolaryngol 4: 213–227

Schuknecht HF (1955) Presbycusis. Laryngoscope 65: 402–419

Schuknecht HF (1964) Further observations on the pathology of presbyacusis. Arch Otolaryngol 80: 369–382

Schuknecht HF (1974) Pathology of the ear. Harvard University Press, Cambridge, Massachusetts

Schuknecht HF, Neff WD, Perlman HB (1951) An experimental study of auditory damage following blows to the head. Ann Otol Rhinol Laryngol 60: 273–289

Šercer A, Krmpotič J (1958) Über die Ursache der progressiven Altersschwerhörigkeit. (Presbyacusis). Acta Otolaryngol (Suppl) 143: 5–36

Součková S (1973) Integration of auditory signals in normal and affected hearing. Cs Otol 22: 267–270 (in Czech)

Spencer JT (1973) Hyperlipoproteinemias in the etiology of inner ear disease. Laryngoscope 83: 639–679

Spencer JT (1975) Hyperlipoproteinemia and inner ear disease. In: Symposium on fluctuant hearing loss, vol 8. Otolaryngol Clinics of North America, pp 483–492

Suga F, Lindsay JR (1976) Histopathological observations of presbycusis. Ann Otol Rhinol Laryngol 85: 169–183

Toynbee J (1849) On the pathology and treatment of the deafness attendant upon old age. Mon J Med Sci, Edinburgh, pp 1–12

von Fieandt H, Saxen A (1937) Pathologie und Klinik der Altersschwerhörigkeit. Acta Otolaryngol (Stockh) (Suppl) 23: 1–85

von Helmholz HLF (1863) Die Lehre von der Tonemfindungen als physiologische Grundlage für die Theorie der Musik. F. Vieweg u Sohn, Braunschweig

Welsh LW, Welsh JJ, Healy MP (1985) Central presbyacusis. Laryngoscope 95: 128–136

Wilde WR (1853) Practical observations on aural surgery and the nature and treatment of diseases of the ear, with illustrations. Churchill, London

Zwaardemaker H (1891) Der Verlust an höhen Tonen mit zunehmendem Alter. Ein neues Gesetz. Archiv für Ohren-Nasen and Kehlkopfheilkunde 32: 53–56

Zwaardemaker H (1894) Der Umfang des Gehörs in verschiedenen Lebensjahren. Z Psychol Physiol 71: 10–28

Clinical Diagnosis and Audiometric Studies

Hearing impairment is common in the elderly. It may be apparent on formal testing even in those subjects in the later years of life who deny any hearing impairment. In this section the clinical diagnosis of the various forms of hearing impairment in the elderly subject will be considered, the established audiometric features of presbyacusis will be reviewed and new audiometric studies of our own will be presented.

Clinical Diagnosis

History

Mode of Onset

In presbyacusis the hearing loss is usually of gradual onset, bilateral and symmetrical. The sudden onset of deafness, particularly in one ear, may be related to viral infection, barotrauma or head injury. Fluctuating hearing loss may be due to fluid in the middle ear, or Ménière's disease. The past history of the patient may give an indication as to the cause of the hearing loss. Enquiry should be made for past noise exposure, whether in military or civilian occupation, past medication with potentially ototoxic drugs, or of general medical illnesses, such as meningitis and syphilis, which can cause damage to the hearing organ.

Tinnitus

Hearing loss in the elderly is often accompanied by buzzing noises in the ear. The noises may be high-or low-pitched and continuous or intermittent (see below). They are frequently louder at night when surroundings are quiet.

Objective tinnitus, which may be heard by the clinician on auscultation, is not a feature of presbyacusis.

Vertigo

The feeling of unsteadiness may be a swaying or rotational one. Acute rotational vertigo implies an end organ lesion. Vertiginous symptoms may be part of the old age disability, but are not a manifestation of presbyacusis.

Ear Discharge

Discharge may be due to infection in the external ear canal or middle ear infection discharging through a tympanic membrane perforation.

Earache

Earache may be due to a local cause in the ear or may be referred from the neck, sinuses or jaw.

Examination of the ear

The pinna and postaural area should be inspected for inflammation, deformity or evidence of previous operation such as depression over the mastoid or a postaural scar. Endaural incisions may be seen anteriorly between the tragus and the rest of helix. A postaural fistula into the mastoid is occasionally seen. Inspection of the ear should be carried out with an electric otoscope after selection of a speculum of suitable size. The pinna should be gently but firmly grasped and pulled posteriorly prior to insertion of the speculum. The tympanic membrane should be examined for perforation, thickening or white chalky patches of tympanosclerosis. Wax is a common problem preventing adequate inspection of the tympanic membrane in the elderly and should be removed by syringing with water at 37°C. Before carrying out this procedure it is advisable to question the patient for symptoms suggestive of perforation of the tympanic membrane.

Clinical Speech Tests

These are useful in order to obtain some information as to how the subject understands the spoken word and is able to discriminate speech. The patient is asked to repeat words spoken to him using a forced whisper and conversational voice at increasing distances from the ear. The ears can be tested separately while a non-tested ear is occluded and masked by pressure on the tragus. In this way an approximation of the degree of hearing loss can be obtained even in the patient's home environment. The examiner will also have obtained some insight into the patient's hearing loss in the course of his history taking. Efforts should now be made to test his hearing, using a forced whisper at different distances from each ear after compressing the tragus of the other ear just firmly enough to close the auditory canal. In this way an approximation of the degree of hearing loss and an idea as to which ear is worse affected can be obtained.

Tuning Fork Methods

A tuning fork at a frequency of 512 Hz is useful in the investigation of elderly patients with hearing disorders. The instrument may be used in the form of two tests:

1. *Weber Test.* The tuning fork is put into vibration and placed with its base on the midpoint of the vertex of the skull. In a subject with normal hearing the sound of the fork is localized centrally within the head or simultaneously in each ear. In a conductive loss the sound of the tuning fork is heard better in the worse ear, i.e. the side of the conductive hearing defect. In sensorineural hearing loss the sound of the tuning fork is heard better in the good ear.

2. *Rinne Test.* The tuning fork is placed in front of the ear to assess air conduction and then on the mastoid process to assess bone conduction. If air conduction is better than bone conduction then sensorineural hearing loss is likely. Conductive hearing loss results in the bone conduction from the mastoid being superior.

Audiogram

The most accurate mode of assessment of the patient's subjective hearing powers is by means of the audiogram. In this method the minimum loudness in dB at various frequencies in kHz, is plotted. The hearing levels may be measured by air and bone conduction. In the latter case the bone conduction on the contralateral side must be masked to avoid "crossed" hearing. The curves obtained by audiometry often provide an indication as to the aetiological basis of the hearing loss (Fig. 2.1).

In some centres hearing is tested by speech audiometry in which the degree of hearing loss in regard to the spoken word is measured by adjusting levels of output from a tape recorder.

Tympanometry

In this investigation the acoustic impedance or stiffness of the eardrum is measured. Changes in impedance reflect the condition of the middle ear. The auditory canal is blocked with a close fitting plug in connection with a pump with which the air pressure in the canal can be adjusted. By an acoustic method the degree of stiffness is recorded in the form of a tympanogram.

The acoustic stapedial reflex may be used to check for recruitment, which is a feature of sensorineural hearing loss of cochlear origin (see Section 3). When a pure tone signal is applied to the ear the middle ear muscles of both ears contract. Normally this takes place at a sound level of about 60 dB. The presence of the stapedial reflex is an indicator of a healthy middle ear and undamaged neural structures of the reflex arc (n.VIII and n.VII). When

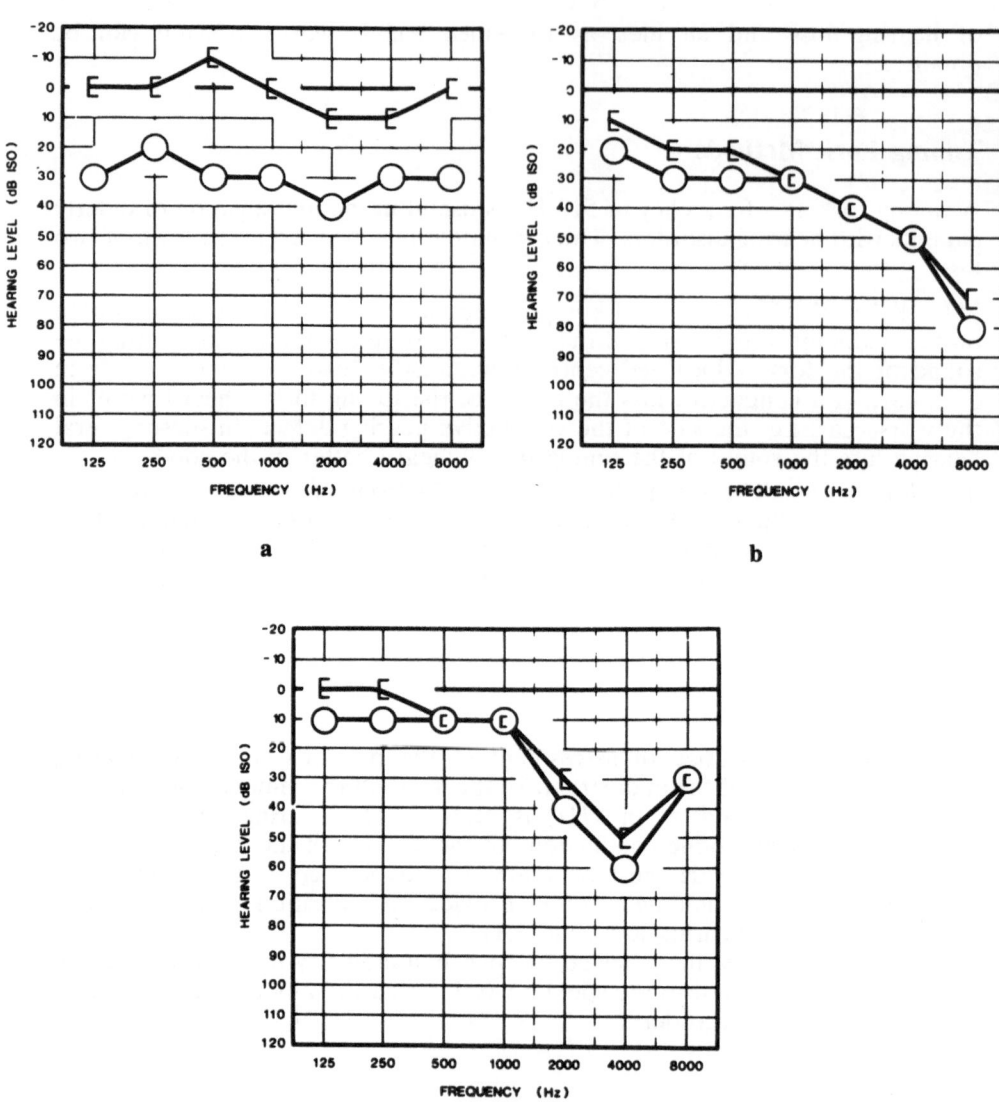

Fig. 2.1a–c. Audiograms from three cases with pathological conditions of the ear. The hearing level in decibels (dB) on the vertical axis is plotted against the frequency of the applied tone on the horizontal axis. In each case bone conduction is shown by a series of *square brackets* and air conduction by a series of *circles*. The normal audiogram gives levels of air and bone conduction between 0 and 10 dB at all frequencies. **a** Conductive hearing loss with air conduction alone somewhat reduced. **b** Old age sensorineural hearing loss with both bone and air conduction steadily diminishing towards higher frequencies. **c** A case of noise induced hearing loss. Here, too, there is a deterioration of hearing in both air and bone conduction towards higher frequencies but there is also a notch at 4000 Hz which is characteristic for this type of hearing loss.

contraction of the ear muscles (stapedius reflex) starts earlier this may point to a process of recruitment.

X-Rays

Plain films of the mastoid may show bone erosion, opacity of the middle ear cleft and a degree of pneumatization. However a much more detailed picture of the middle ear, facial nerve and ossicles can be obtained by hypocycloidal polytomography. Computerized tomogram scanning with enhancement is useful in the diagnosis of small acoustic neuromas which may present as old age hearing loss.

Electrophysiology

These tests are useful because they are objective, i.e. they do not require the positive co-operation of the patient. This is important in the elderly patient with even minor degrees of dementia which may affect the carrying out of adequate history taking, physical examination and subjective audiometry. Acoustic stimulation may be measured from the brainstem (brainstem evoked responses), and from the cochlea (electrocochleograms) (see Section 3).

Calorics and Electronystagmography

These are tests of the balancing mechanism and measure responses of the labyrinth to caloric and rotatory stimuli. They record the activity of the integrity of the vestibule and related structures of the inner ear, brainstem and higher centres related to the balance mechanism, including eye movements. Changes in these regions may accompany old age hearing loss and their presence may assist diagnosis of the cause of the condition.

Disorders of the External Ear

Otitis externa is very common in the elderly, and may give rise to conductive hearing loss by constricting the external auditory meatus. On the pinna frostbite, chondrodermatitis, nodularis helicis (a painful nodule on the helix), and two forms of neoplasm, squamous cell carcinoma and basal cell carcinoma, may be seen. In the external auditory meatus inflammatory conditions due to viruses (particularly herpes zoster and simplex), bacteria (especially *Staphylococcus aureus* and *Pseudomonas aeruginosa*) and fungi (particularly strains of *Aspergillus*), are also common. Bony osteomata or exostoses may give rise to hearing impairment and are usually associated with cold water swimming in earlier life.

Attention has been drawn to the existence of minor disturbances of the external auditory meatus in the elderly which may lead to difficulties in the fitting of hearing aids for the rehabilitation of hearing loss (Maurer and Rupp 1979). Two such conditions are particularly important:

1. The most common of these conditions is the presence of a large amount of hardened wax in the external auditory meatus which may become impacted and lead to conductive hearing loss. It is possible that the reduced tactile sensation that is said to occur in elderly people may result in the subjects not being able to feel the presence of accumulation of wax in the external auditory meatus.

2. The other condition is one of prolapsed ear canal caused by loss of elasticity of the cartilage of external auditory meatus resulting in a partial, or in rare cases a complete, dropping of the canal. This may lead to a conduction type hearing loss and may also cause difficulties in the fitting of moulds for hearing aids.

Disorders of the Middle Ear

Chronic suppurative otitis media is a not infrequent finding in the elderly. There may be perforation of the tympanic membrane, polyp formation, tympanosclerosis or cholesteatoma and operative treatment may be required. The characteristic audiogram feature is reduction of hearing perceived through air conduction, with normal hearing by bone conduction (Fig. 2.1).

The middle ear is said to manifest a special form of ageing activity with involvement of the joints between the ossicles by arthritic changes and eventual ankylosis (see Section 1). This has been stated to produce in all elderly people a mild degree of hearing loss. However in an examination of bone and air conduction in approximately 100 elderly subjects in a geriatric hospital we were not able to find any widening of the air–bone gap of any significance. In an additional study of 25 elderly people for impedance changes we were not able to detect such changes, except in the few cases that showed features of mild otitis media.

Disorders of the Inner Ear

Presbyacusis

The commonest cause of deafness in the elderly is presbyacusis, which is dealt with at length in this monograph.

The presenting symptom is usually hearing loss which may be accompanied by tinnitus. The patient may state that the loss is particularly severe for high tones. Middle ear and central nervous system pathology should be excluded by appropriate clinical examinations. Tuning fork and audiogram investigations confirm the hearing loss, particularly at high frequencies.

The patient with presbyacusis may complain of discomfort on hearing a loud noise and this may indicate recruitment. This is a subjective phenomenon in which an ear with sensorineural hearing loss such as in presbyacusis seems to hear tones to a louder degree than the normal ear. Recruitment may be confirmed by the short increment sensitivity index (SISI) test, in which a short increment of intensity imposed on a carrier tone is used. A carrier tone starting at 20 dB is given 20 increments of 1 dB, each lasting for a fraction of a second. Normal ears will hear only about 20% of the increments. Subjects with presbyacusis usually hear almost 100% of the increments, indicating a high degree of recruitment and therefore of cochlear damage. Many elderly patients may not be able to concentrate sufficiently for this test. In doubtful cases brainstem evoked responses may assist in the diagnosis (see Section 3).

Noise Induced Hearing Loss

Awareness of the possibility of noise as the basis of hearing loss in an elderly patient will usually come about in the history-taking. The audiogram characteristically shows a notch in the higher tones (Fig. 2.1).

Ototoxicity

Damage to the cochlea from ototoxic drugs is also usually manifested by a sensorineural hearing loss with no other clinical manifestations. Again the history of the patient must be taken carefully to provide evidence for this lesion.

Bone Lesions of the Otic Capsule

Paget's disease of bone frequently affects the bony labyrinth. In some of these cases sensorineural deafness may be the presenting clinical symptom. Examination will reveal the characteristic osseous changes of Paget's which will be confirmed by radiological investigation, and there is elevation of the serum alkaline phosphatase.

Otosclerosis may present as sensorineural deafness in an elderly subject. This is thought to be the result of involvement of the cochlea by the otosclerotic process. Evidence of conductive deafness will also be found on audiometry and radiological features of otosclerosis will be present.

Acoustic Neuroma

A schwannoma of the eighth nerve may be the cause of hearing impairment in a small proportion of elderly patients. There may be concomitant vestibular symptoms such as dizziness. Nystagmus and alteration of the caloric responses will usually support the diagnosis, which will be finalized on the basis of computerized tomogram scans of the internal auditory meatus.

Tinnitus

We have found that 20% of people over 65 years of age have tinnitus. It is usually intermittent and only one in twenty-five are seriously troubled by it.

The cause of tinnitus in the aged is not known, but is thought to be associated with hair cell or neuronal degeneration.

Elderly patients with severe tinnitus find that the noise is worse at night when surroundings are quiet and less during the day when there is more extraneous sound. This masking effect of external noise may be used by the clinician in treatment, several different devices being available for this purpose.

Audiometric studies

An important special investigation of any type of hearing loss is to assess the degree of disability of the sufferers as measured by the most widely standardized test available, pure tone audiometry. In the remainder of this section the findings resulting from the investigation of the hearing of elderly people by this method are presented.

Numerous studies have been published on the changes which occur in the pure tone threshold as a function of age. Each has employed its own criteria for the selection of subjects. Different age ranges have been considered (Table 2.1). Some studies excluded, for example, subjects with an ear disorder, a history of noise exposure, differences in thresholds between the ears greater than a certain value, and a low frequency hearing loss; others included all subjects irrespective of noise exposure and otological disorder.

Few studies have considered the deterioration of hearing beyond the age of 70 years or, if they have done so, the sample has been small. Thus in their study by Møller (1981) indicated that hearing thresholds in women deteriorated on only 54 ears over the age of 70 years; the mean age was 74.5 years. Kell and his colleagues (1970) from a group of 188 males and 238 females between the ages of 20 and 80 years had only 23 males and 33 females beyond the age of 65 years.

As life expectancy has increased, interest in the auditory well-being of older people has grown. One of the earliest studies on people beyond 70 years was that by Sataloff and Manduke (1957) who indicated that the hearing ceased to deteriorate after the age of 65 years. However, other studies such as those by Goetzinger and his colleagues (1961) and Milne and Lauder (1975) have shown continuous deterioration in hearing acuity with increasing age. Although a study by Møller (1981) indicated that hearing thresholds in women deteriorated throughout the entire frequency range over the 5-year period from age 70 to 75 years, no detectable change in thresholds could be found over the same period for men. This might suggest on the one hand that a plateau could be reached once the subject's hearing had attained a certain level. On the other hand it might merely reflect that we are dealing with a survival population, those who survive in later years being not only generally fit, but specifically so in relation to their hearing. Of course, both hypotheses may be true.

There is thus a need for a study to resolve these differences by drawing on larger numbers of people at more advanced ages then those described in the literature. With this in mind we set out to investigate the hearing function of in-patients in medical wards for elderly people, using pure tone audiometry as the main tool. The following studies have not yet been reported and are

Table 2.1. Ten studies of hearing threshold as a function of age

Reference	Year	Number of subject	Age range	Measure and dispersion	Types of subjects
Sataloff and Manduke	1957	94 males 101 females	64–91 64–93	Mean value for better ear	Various, screened
Jatho and Heck	1959	351 males	15–75	Mean, range	Not stated
Hinchcliffe	1959	161 males 157 females	21.5–70 21.5–70	Median, Q1Q3	Random, unscreened
Goetzinger et al.	1961	45 males	60–80	Mean + SD	Homes for the aged, screened
Glorig and Nixon	1962	2518 males (right ears)	20–79	Median, none	Professional, unscreened
Corso	1963	247 males 377 females	21.5–62.5 21.5–62.5	Mean + SD Median	Random, unscreened
Taylor et al.	1967	171 females	22.4–65	Mean + SD	Teaching staff, screened
Kell	1970	188 males 238 females	20–80 20–80	Mean + SD	Orkney population, screened
Milne and Lauder	1975	215 males 272 females	62–90 62–90	Median 95% confidence limits	Random NHS lists, unscreened
Møller	1981	179 males 197 females	70	Mean + SD	Random, unscreened

SD: standard deviation.
Screened: middle ear disease and acoustic trauma excluded.
Unscreened: middle ear disease and acoustic trauma not excluded.
NHS: National Health Service.

presented here as the results throw light on some of the problems of old age hearing loss.

Personal Investigations

Hearing Levels in a Large Group of Elderly People

Procedure

The subjects were in-patients comprising people in the age range 60 to 94 years. A total of 169 patients was examined. The average age for men and women tested was 79.7 (SD 7.42) and 81.5 (SD 6.79) years respectively. The subjects suffered from a variety of medical conditions. The reasons for admission ranged from medical factors to social ones. Some were acute cases admitted for rehabilitation.

Hearing thresholds were measured using an Amplivox model 84 audiometer with TDH 39 headphones. Calibration to British Standards was performed every 6 months throughout the period of study. As no accessible room within the hospital had acceptable ambient noise levels and, since a high proportion of patients had poor mobility and there were no sound treated rooms in the hospital, the patients were tested at the bedside. Steps were taken to minimize the background noise by turning off all radio and television sets on the ward, closing doors and pulling curtains around the patient.

To determine the possible detrimental effects that the background noise had on the measured threshold, a group of young people, between the ages of 19 and 25 years, was tested under the same conditions. None of these subjects had any auditory symptoms and none had any otoscopic abnormality. The mean thresholds and the standard deviations for the 12 young subjects are shown in Fig. 2.2. The background noise was seen to influence the measured thresholds. The degree of masking appeared to change from above 5 dB over the frequency range from 2 kHz to 4 kHz and down to about 15 dB at 250 Hz to 125 Hz. Allowing for the dispersion of the data it would appear that this system for measuring hearing is not able to measure thresholds of hearing better than about 10 dB for the frequencies of 2 kHz and above and better then 20 dB for frequencies of 1 kHz and below.

Air conduction audiometry was carried out using the abridged ascending procedure as described by the consultative document on International Standards Organisation (ISO) standards for audiometry (DIS 8253). The procedure to be used was first explained to the patient. Since many subjects suffered from arthritis or a previous stroke they were asked to respond verbally to the test tone. Unless the patient considered the hearing to be better in the left ear, the right ear was tested first. The order in which the threshold was measured at the various frequencies was 1, 2, 3, 4, 6, 8, 0.5 and 0.25 kHz. Finally the threshold at 1 kHz was re-measured to check for reproducibility. Thirty-one per cent of the patients were retested with an average interval of one week between test and retest. The majority of patients (82%) had a mean difference of ±5 dB or

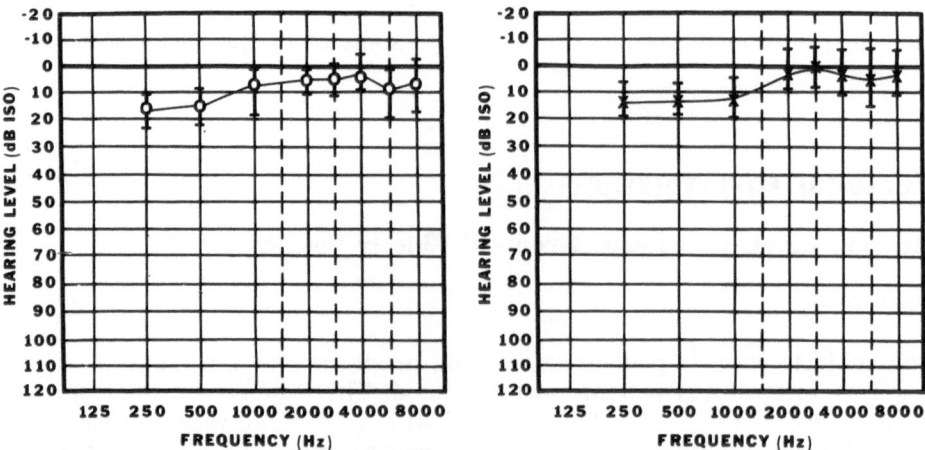

Fig. 2.2. Mean thresholds with standard deviations for 12 young subjects under ambient conditions of investigation. (○) right; (×) left.

less between test and retest. For the other patients, it was noted that the large variation in threshold occurred at the low frequencies 250 and 500 Hz, where background noise particularly influences hearing thresholds.

Results

A positively skewed distibution of the hearing levels at the higher frequencies was found. The median values together with the lower and upper quartiles

Table 2.2. Median values of hearing loss (dBHTL) for right ear in elderly men

Frequency (Hz)	Age groups (mean age + SD)			
	60–69[a] (65 ± 3.85)	70–79[b] (74.41 ± 3.10)	80–89[c] (83.37 ± 2.79)	90–99[d] (91.00 ± 1.07)
250	22.5 25.0 32.5	25.0 32.5 40.0	30.0 37.5 60.0	20.0 40.0 80.0
500	20.0 25.0 37.5	30.0 35.0 40.0	30.0 45.0 56.3	25.0 40.0 75.0
1000	22.5 23.0 37.5	25.0 35.0 51.3	30.0 40.0 61.3	30.0 55.0 65.0
2000	17.5 25.0 45.0	25.0 42.5 65.0	33.8 55.0 66.3	50.0 60.0 80.0
3000	22.5 35.0 65.0	42.5 60.0 72.5	45.0 65.0 75.0	60.0 72.5 101.3
4000	32.5 55.0 70.0 M5	48.7 65.0 81.3 M5	60.0 80.0 85.0	65.0 90.0 110.0
6000	50.0 75.0 80.0 M1	60.0 75.0 85.0	70.0 80.0 92.5	95.0 NR NR M5
8000	32.5 55.0 65.0	62.5 80.0 NR	72.5 90.0 NR	91.3 NR NR

The central figure in each case is the median, the upper and lower figures being the 75% and 25% percentile levels.
NR: no response.
Kruskal–Wallis:
  M1: The difference between the medians of the men and women is significant at 1% level.
  M5: The difference between the medians of the men and women is significant at 5% level.
[a] $n = 5$.
[b] $n = 22$.
[c] $n = 30$.
[d] $n = 7$.

were therefore selected to describe the data rather than the mean values, although when considering hearing losses with individual frequencies in a later part of the investigation mean values were used. Non-parametric measures (median together with the upper and lower quartiles) of the measured thresholds of hearing for the two ears, two sexes and various age groups are shown in Tables 2.2, 2.3, 2.4 and 2.5.

General Analysis. Examination of the median values for four decades indicates that hearing deteriorates with age. This was confirmed by the Kruskal–Wallis non-parametric one-way analysis of variance (Milton and Tsokos 1983). The hearing loss in persons of 90 years and over was found to be significantly greater than that for people of 60–69 years of age at all frequencies and for both sexes except for frequencies below 1 kHz on the right and 3 kHz on the

Table 2.3. Median values of hearing loss (dBHTL) for left ear in elderly men

Frequency (Hz)	Age groups			
	60–69	70–79	80–89	90–99
250	25.0	30.0	25.0	35.0
	30.0	35.0	40.0	40.0
	55.0	51.3	45.0	80.0
		M5		
500	27.5	30.0	30.0	35.0
	35.0	40.0	40.0	55.0
	42.5	55.0	55.0	75.0
	M5			
1000	20.0	25.0	30.0	20.0
	25.0	35.0	40.0	40.0
	37.5	70.0	56.3	70.0
		M5		
2000	22.5	28.8	40.0	40.0
	35.0	52.5	55.0	50.0
	50.0	71.3	60.0	75.0
3000	32.5	41.3	60.0	55.0
	45.0	62.5	65.0	72.5
	57.5	75.0	80.0	88.8
	M5	M1		
4000	30.0	60.0	65.0	65.0
	45.0	70.0	72.5	75.0
	70.0	86.3	86.3	95.0
		M1		
6000	45.0	60.0	70.0	87.5
	70.0	75.0	85.0	NR
	80.0	87.5	NR	NR
	M5	M5		
8000	60.0	67.5	78.7	91.3
	65.0	80.0	90.0	NR
	70.0	NR	NR	NR
	M1	M1	M5	

For Key, see Table 2.2.

Table 2.4. Median values of hearing loss (dBHTL) for right ear in elderly women

Frequency (Hz)	Age groups (mean age + SD)			
	60–69[a] (66.43 ± 2.19)	70–79[b] (75.50 ± 2.16)	80–89[c] (84.72 ± 2.52)	90–99[d] (91.18 ± 1.27)
250	10.0	28.8	30.0	40.0
	20.0	35.0	40.0	45.0
	25.0	40.0	55.0	60.0
500	20.0	30.0	35.0	50.0
	20.0	35.0	45.0	60.0
	30.0	51.3	62.5	70.0
1000	20.0	25.0	35.0	45.0
	20.0	35.0	45.0	60.0
	35.0	50.0	65.0	70.0
2000	20.0	28.8	40.0	45.0
	25.0	40.0	55.0	60.0
	35.0	50.0	70.0	65.0
3000	18.8	41.3	46.3	62.5
	25.0	47.5	60.0	75.0
	27.5	60.0	75.0	80.0
4000	15.0	43.8	55.0	65.0
	20.0	52.5	65.0	70.0
	35.0	66.3	85.0	80.0
6000	15.0	53.8	65.0	73.8
	25.0	60.0	80.0	85.0
	33.0	80.0	88.8	92.5
8000	15.0	60.0	72.5	85.0
	50.0	73.0	85.0	90.0
	50.0	85.0	NR	NR

For Key, see Table 2.2.
[a] $n = 7$.
[b] $n = 30$.
[c] $n = 57$.
[d] $n = 11$.

left for men. A significant increase in hearing loss was noticed between successive age groups in females except between the eighth and ninth decade at 3 kHz ($p < 0.01$) and 6 kHz ($p < 0.04$) both on the left side.

Comparison of Hearing Loss in Right and Left Ears. Differences between the left and right hearing thresholds were investigated using the non-parametric Wilcoxon matched-pairs signed-ranks test. This test showed that there was no significant difference between the left and right ear.

Comparison of Hearing Loss in the Sexes. The possibility of hearing loss differences between the sexes was also examined with Kruskal–Wallis non-parametric one-way analysis of variance (Milton and Tsokos 1983). The test showed 15 differences out of 64 measurements undertaken. These differences are particularly noticeable within the first two decades (70–79 years, 80–89 years). In the higher frequencies in particular (3–8 kHz) the men had the greater hearing loss (see Tables 2.2 and 2.3).

Table 2.5. Median values of hearing loss (dBHTL) for left ear in elderly women

Frequency (Hz)	Age groups			
	60–69	70–79	80–89	90–99
250	20.0	18.8	30.0	30.0
	20.0	30.0	33.0	45.0
	30.0	40.0	53.8	55.0
500	20.0	23.8	35.0	40.0
	25.0	32.5	45.0	50.0
	30.0	41.3	55.0	63.0
1000	20.0	20.0	33.0	40.0
	20.0	32.5	45.0	53.0
	25.0	40.0	62.5	70.0
2000	10.0	25.0	40.0	50.0
	25.0	40.0	50.0	55.0
	25.0	45.0	65.0	65.0
3000	10.0	33.0	50.0	65.0
	15.0	45.0	60.0	70.0
	30.0	55.0	70.0	75.0
4000	20.0	35.0	60.0	65.0
	23.0	52.5	65.0	70.0
	45.0	60.0	73.0	75.0
6000	35.0	53.8	70.0	77.5
	35.0	65.0	80.0	90.0
	41.3	71.3	88.8	NR
8000	40.0	60.0	70.0	85.0
	40.0	67.5	80.0	NR
	30.0	73.0	NR	NR

For Key, see Table 2.2.

General Conclusion.   The median hearing losses for males and females are depicted in Fig. 2.3 and 2.4 respectively. Consideration of these figures and Tables 2.2 to 2.5 shows that the hearing loss was greater in men compared with women of all ages for the higher frequencies (2 kHz and above) and greater in older compared to younger persons, for both sexes.

Hearing Loss at Different Frequencies.   Collections of published data relating the elevation of hearing threshold levels to increasing age at different frequencies are available. They show some variations caused by different criteria for selection of tests and audiometric techniques used. For this reason a representative set of values has been established on an international level referring to a large screened population of otologically normal persons: this is known as the International Standard (ISO 1984). This set is composed of the statistics of the hearing levels of populations of various ages. The data can also be used for comparison of an individual's hearing with the normal distribution of the hearing threshold levels within the person's age group. The International Standard document specifies the expected value of the median hearing threshold shift in comparison with a group of persons of 18 years of age. It also

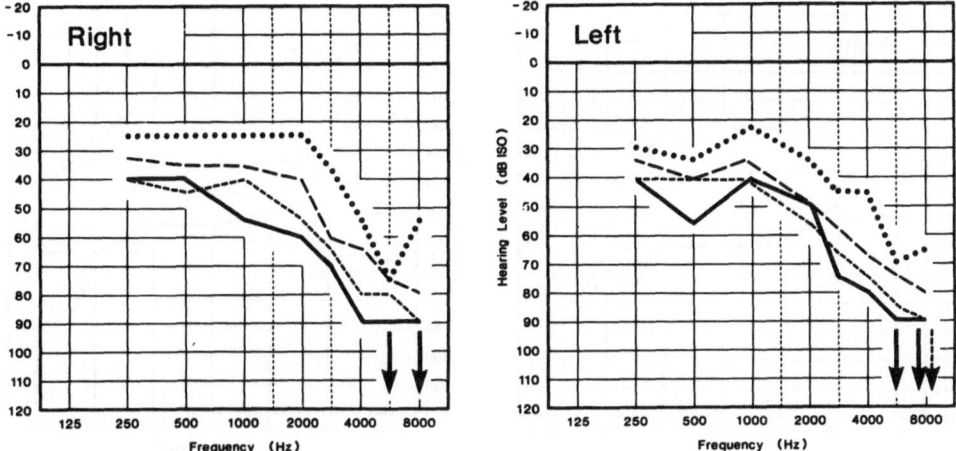

Fig. 2.3. Median audiometric hearing losses of 64 men in decades at frequencies 250 Hz, 500 Hz and 1, 2, 3, 4, 6 and 8 kHz. (....) 60–69 years, $n=5$; (– – – –) 70–79 years, $n=22$; (----) 80–89 years, $n=30$; (——) 90–99 years, $n=7$.

specifies the expected statistical distribution above and below the median value for the range of audiometric frequencies from 125 Hz to 8 kHz and for groups of otologically normal persons of a given age within the age limits of 18–70 years inclusive.

In the present audiometric study we attempted to line up the values in elderly patients of the median hearing threshold shifts for each frequency and for each age group, for both men and women, with the ISO 7029 curves. First, figures of pure tone audiometric thresholds in the ISO 7029 data (median values) were plotted on the y axis against age on the x axis. We obtained thus a graph for the lower range of ages, 20–70 years. Median values of hearing threshold levels for elderly patients were then plotted for 65, 74.4, 83.4 and 91 years of age at each frequency using our own data given in Tables 2.2, 2.3, 2.4 and 2.5. An approximately S-shaped curve was obtained using the above data with an interval at about the midpoint (Fig. 2.5). A to B represents ISO 7029 figures. From B to C there is an interruption of the S-shaped curve. C to D represents the median audiometric values from the geriatric unit obtained by us. We tried to calculate the third degree of polynomial equation to fit the data in order to close the gap and obtain a single curve, but were not able to do this to our satisfaction.

We then tried to tabulate the values for the elderly directly by the equation provided by the ISO document for the higher age range according to the formula given and the appropriate coefficient. In this way, we obtained values for men and women. These were the values calculated by extending the age

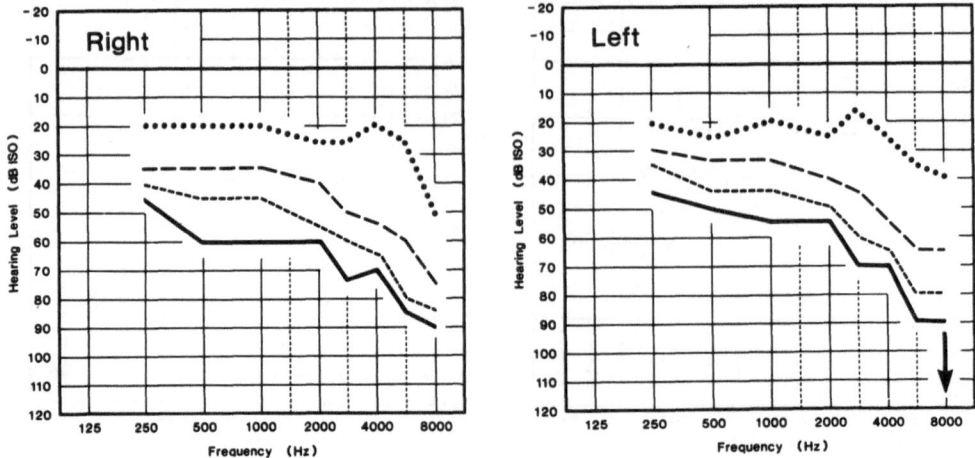

Fig. 2.4. Median audiometric hearing losses of 105 women in decades at frequencies of 250 Hz, 500 Hz and 1, 2, 3, 4, 6 and 8 kHz (. . . .) 60–69 years, n=7; (– – – –) 70–79 years, n=30; (----) 80–89 years, n=51; (——) 90–99 years, n=11.

range beyond those given in the document. The results were impossibly low for right and left median hearing losses for the higher frequencies and in the higher age range. It seems that the equation given by ISO 7029 is not valid under these circumstances.

The data on which ISO 7029 is based are sparse above 60 years, for the reason that it is hard to find sufficient numbers of people at this age who qualify for the description "otologically normal". Most people suffer some kind of auditory deficit by the time they reach 70 years of age. ISO 7029 should not, therefore, be pushed beyond its already stretched limits.

We then tried to line up the data with those collected by Robinson (1988) from median values for a typical unselected, but non-clinical population, derived from surveys, mostly American, involving 30 000 or more people of all ages up to about 70 years (Fig. 2.6). There is reasonable agreement of the data with regard to both groups of men in the range 65–70 years at 6, 4 and 3 kHz. Both groups have excessive loss at 2 and 1 kHz. The 8 kHz frequency was not tested in any of the large scale studies that Robinson examined and therefore not considered. There is usually an increase of hearing loss with the frequency increase, but the 2 kHz frequency curve levels off and the slope appears to be asymptotic beyond 80 years. This is not shown at other frequencies.

In women the hearing loss steadily increases until 90 years of age (Fig. 2.7). The slope appears to be asymptotic with ages beyond 90 years at 2 kHz. This again is not apparent at other frequencies, as with men.

Attempts have been made in the past to extrapolate hearing levels against age so as to quantify the audiological features of ageing (Spoor 1967). Spoor's

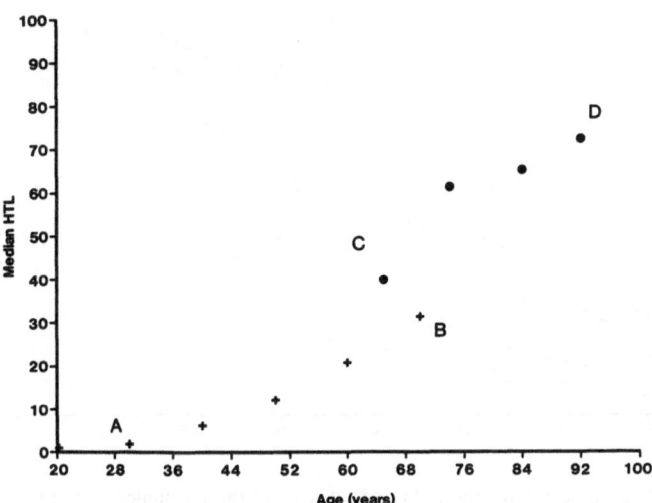

Fig. 2.5. Median values of hearing losses in men at 3 kHz in the present study extrapolated into median values of ISO document (1984). For explanation, see text.

database was extracted from a number of studies and covered a wide range from 18 to 75 years of age. The collected data were then extrapolated to beyond 75 years by using a logistic function for his calculations. By introducing three constants to his equation he showed that hearing loss increased exponentially as a function of age starting from the age of 25. In our sample (Figs. 2.6 and 2.7) for both sexes with presbyacusis, the hearing losses were higher than those in Spoor's data. It seems that logistic functions do not represent the numbers well for the higher age range.

"Screened" and "Unscreened" Populations Compared

The significance of noise and other agents deleterious to hearing in the production of the specific hearing loss in the aged was questioned. To obtain information on this point separate curves were drawn at each frequency for groups of individuals who were found to lack evidence of severe noise exposure or other possible otological disturbance. The thresholds so obtained were compared with previously described thresholds for the total elderly population depicted, without regard to noise or otological damage. The thresholds for the screened population are shown in Fig. 2.6 and 2.7.

In men aged 70–79 years the screened groups of subjects presented almost identical curves at each frequency charted from 1 kHz until 6 kHz. A slight difference, to the amount of 2.5 dB, was noted at 2 kHz in subjects over 90 years old. For the women the thresholds for the groups of screened subjects aged 65 and over were also almost identical to the unscreened groups in all

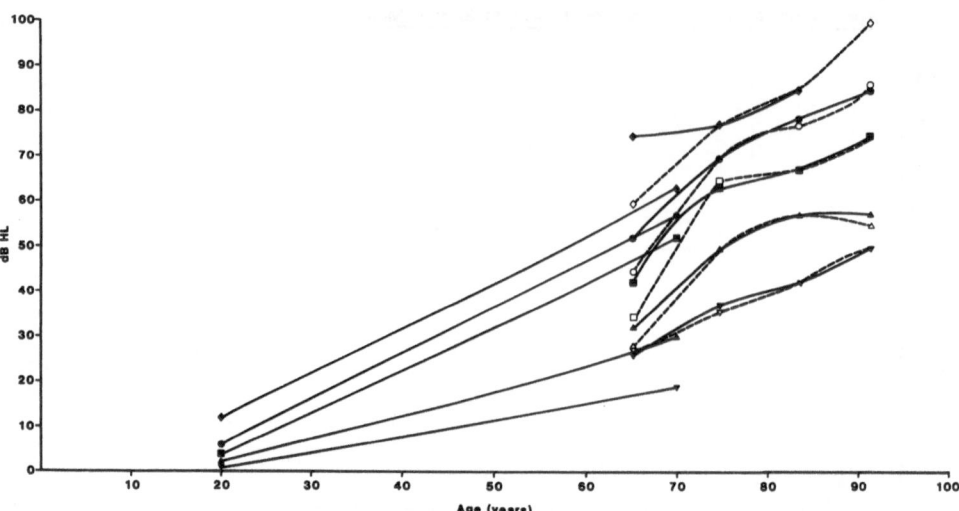

Fig. 2.6. Average of median hearing losses for the right and left ears at the frequencies of 1(▼), 2(▲), 3(■), 4(●) and 6(◆) kHz in elderly men. These values (right side of figure) are extrapolated into a database derived from a non-clinical population from surveys involving 30 000 people up to 70 years of age (left side of figure). The *full lines* on the right represent subjects unscreened for otological problems and noise exposure. The *broken lines* on the right show the population of screened men (no noise exposure and no otological problems).

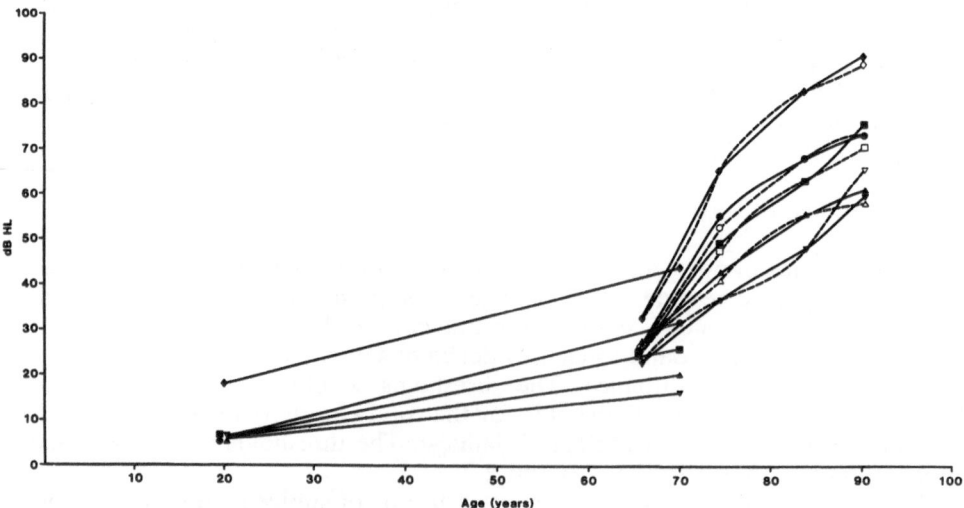

Fig. 2.7. Average of median hearing losses for the right and left ears at the frequencies of 1(▼), 2(▲), 3(■), 4(●) and 6(◆) kHz in elderly women. The *full lines* on the right represent subjects unscreened for otological problems and noise exposure. The *broken lines* on the right show the population of screened women (no noise exposure and no otological problems).

curves from 1 kHz until 6 kHz. Slight diversions were present to the extent of 2.5 dB at 2 kHz for 91 years and older and 2.5 dB at 4 kHz for the 75.5 years group.

Discussion

Recent studies by Milne and Lauder (1975) and by Møller (1981), which concentrated on hearing in the elderly population, used unscreened subjects. The earlier published studies, however, relating hearing loss as a function of age were performed in selected populations, excluding subjects who had any audiological disorders, asymmetrical hearing loss and noise exposure. Differences in populations between studies prevent fair comparison being made. Of the studies that used unscreened populations, Møller (1981) used only people over 70 years old for her cross-sectional study and Milne and Lauder (1975), whose age range lies within this present study, used different age group sizes and do not state the average age within each group. These two factors prevent easy comparison being made.

Milne and Lauder (1975) were unable to find a significant difference between the left and right ear. This must inevitably result in a more favourable picture of the hearing threshold in the elderly, since the better ear is used for analysis in individual cases. Other studies have noticed that the mean hearing loss is greater for the left than the right ear above 2 kHz. Thus Glorig (1957), Hinchcliffe (1959), and Goetzinger et al. (1961) found that hearing levels in the right ear were significantly better than in the left ear, when frequency was disregarded. The present study therefore analyses the results for each ear separately. No statistically significant difference was found between the left and right ears in the present study. This is in agreement with the findings of Møller (1981).

The values for hearing thresholds in women found in this study are in fairly close agreement with those of Milne and Lauder (1975). The hearing thresholds for men are slightly higher than those published by Milne and Lauder (1975). The interquartile range at each frequency within any group is quite large, which may reflect many factors that may influence hearing. The age effect on hearing was found to be present in both men and women at high frequencies. This effect was also noticed at low frequencies for females. However the results for low frequencies must be interpreted cautiously because of the level of background noise. On examining the median values at 250 and 500 Hz it was noticeable that females had worse hearing acuity than men, except in the age group 60–69 years. This has been reported by others (Goetzinger et al. 1961; Kell et al. 1970; Milne and Lauder 1975; Milne 1977).

It was noticeable in women that the hearing deteriorated between each decade (Fig. 2.3). This was significant except between the eighth and ninth decade. The women aged 90 and over were all in their early nineties (mean age of 91.2 years); this could explain why no statistical difference from the group aged 80 and over was seen. Although an age effect was also present in men, the deterioration was not so marked between decades as with women. The range of hearing threshold in each decade was larger for men than that for women; so stated earlier the male population is less likely to be homogeneous in comparison to the female population with regard to noise exposure. The majority of women in this study had spent most of their lives as housewives or in occupations where noise levels were low.

Corso (1963) found that the rate of deterioration in women was fairly uniform as a function of age. However for the men the rate was found to very with age in a discontinuous manner and marked changes in hearing occurred in men every 15 years on average. Møller (1981) also noticed a change in rate of deterioration in hearing acuity between men and women over a 5-year period as stated earlier. It would appear, therefore, that both sexes have an age effect but the rate of deterioration in females compared with males may differ. Only a longitudinal study of long duration could verify such a conclusion.

A difference between sexes was found in this study, in agreement with other studies (Hinchcliffe 1959; Goetzinger et al. 1961; Corso 1963), in that the hearing loss in males exceeds that for women for frequencies at and above 2 kHz. A significant difference was found at some frequencies, especially in the left ear for the sixth and seventh decades as shown in Table 2.3. In the frequencies below or at 1 kHz, men supposedly have better hearing. There was no significant difference to support this supposition and in fact the converse was found for the left ear at 500 Hz in the sixth decade and 250 and 1 kHz in the seventh decade in this study. One must remember the level of background noise affecting the low frequencies when evaluating the latter findings. When interpreting the sex difference we must question whether it is the habits and occupational variations between the sexes which produce this effect. The lesser hearing loss in females compared to males may be a result of less exposure to noise during their lives. When questioned about noise exposure 41% of women and 64% of men claimed to have worked in a noisy environment at some time during their life. The percentages of subjects associating a noisy environment with a particular occupation are recorded in Table 2.6.

The degree of noise exposure is very difficult to assess, as each patient's tolerance to noise is different. The 31% of the men recorded to be in the forces were actively engaged in fighting during the First or Second World Wars and therefore experienced gun firing.

Although in each decade there is a wide variation in hearing acuity, a result perhaps of the differences between patients' exposure to noise and otological disorders, an overall deterioration was found between the sixth and ninth, decade, as shown by Fig. 2.2 and 2.3. One must conclude from this that hearing acuity seems to degenerate progressively throughout later life.

The effect of screening the subjects for otological disorders and noise history had a negligible result in producing alterations of the hearing loss curves at the different frequencies tested. It has already been suggested that the noise

Table 2.6. Percentages of subjects with history of exposure to various noisy occupations

Women	%	Men	%
Office work	7	Forces	31
Kitchen	6	Factory work	18
Dressmaking	7	Construction work	6
Factory work	17	Music	5
Various	4	Various	4

exposure of the subject may not be a factor contributing seriously to the amount of hearing loss in old age, the greatest effect of noise being in the early years after exposure to it (Glorig and Davis 1961). Our observations are also in keeping with the concept that there is a time in later life beyond which earlier damage no longer plays a part in aggravating the hearing loss.

Summary

An investigation of the hearing function of elderly people was undertaken on 169 patients using pure tone audiometry. The average age for men and women tested was 79.69 years (SD 7.42) and 81.54 years (SD 6.79), respectively. The non-parametric Wilcoxon matched-pairs signed-ranks test showed no significant difference between the left and the right ear in the patients. Analysis of the median values for the decades 60–69, 70–79, 80–89, 90–99 years by Kruskal–Wallis one-way analysis of variance showed a hearing loss in the group of 90 years and over at the 1% level, greater than in the decade of 60–69 years, at all frequencies for both men and women except for frequencies below 1 kHz on the right and below 3 kHz on the left for men. A significant increase in hearing loss was found between the successive age groups in women except between the eighth and ninth decades. Higher frequencies of 2 kHz and above showed a difference between the sexes, men having a greater loss.

With regard to the left ear, the decade 70–79 showed the most significant loss of hearing in comparison with that of 60–69 years. Our data for hearing losses in elderly men at frequencies of 1, 2, 3, 4, and 6 kHz are in reasonable agreement with those of a large pooled group of men at the same ages who were examined by Robinson (1988).

When hearing losses at the different frequencies of the group as a whole are compared in men and women separately with hearing losses of men and women who have been screened for otological disorders and excessive noise exposure, those with a positive history of the latter being eliminated, insignificant differences only were found.

Comparison of Hearing Levels in Different Medical and Environmental Conditions

It has been suggested that the hearing level in old age may be depressed by the presence of serious general disease such as myocardial infarction or hypertension (Rosen et al. 1964). To test this we carried out an audiometric study on four samples of elderly people in different medical and environmental conditions.

Materials and Methods

The samples consisted of four groups of people aged over 65 years:

1. Elderly people visited at home by medical personnel for investigation of deafness (domiciliary visits).
2. Patients presenting at a hearing aid clinic.
3. Elderly people living in residential homes for the elderly.

4. Hospital in-patients who had been admitted for many different conditions.

Environmental Conditions in the Four Groups

1. Domiciliary Visits.   This group was questioned and examined in their own homes. They were often not bedridden. Ambient noise levels in these private houses and flats were usually acceptable for audiometry.

2. Hearing Aid Clinic.   In this group audiometry was carried out in sound-proof rooms in the centre. The patients were referred by their general practitioners or ENT surgeons for a hearing aid fitting. The majority were fit to walk.

3. Residential Homes for the Elderly.   Five such homes were visited. The room chosen for investigation was usually the "surgery" where the elderly people were normally cared for by the general practitioner or nursing staff. The environment was acceptable for audiometry as regards ambient noise.

4. Hospital.   As no accessible room within the hospital had acceptable ambient noise levels, similar measures were undertaken to those described above for the audiometric study of patients in a hospital environment.

Instrumentation

To minimize the effects of ambient noise in the lower frequency ranges of 250 Hz and 500 Hz, Audiocups, which are enclosures for audiometer headphones with exceptional noise-excluding properties (obtained from P.C. Werth Ltd, Audiology House, 45 Nightingale Lane, London SW12) were fitted to TDH 39 headphones for investigation of subjects in the hospital. An Amplivox model 84 portable diagnostic audiometer was used throughout this study, except for the patients who presented at the hearing aid clinic where a Grason Staedler model was used. The audiometer was calibrated in accordance with British standard procedure every 6 months throughout the testing period and was tested by the operator at the commencement of each day.

Results

Means were used in this part of the investigation rather than medians as a description of the hearing losses, to bring them into line with the findings of other studies of hearing loss under different environmental conditions in the literature. The oldest group was constituted by those in the residential homes at a mean of 85 years (SD 7.5) for the 106 residents. The youngest was that of hearing aid clinic at a mean age of 76.4 years (SD 6.5). The age and sex distribution of the four groups is shown in Fig. 2.8 and 2.9.

Fig. 2.10 shows the mean hearing losses at frequencies of 250 Hz, 500 Hz, 1 kHz, 2 kHz, 4 kHz and 8 kHz for hearing in both ears combined. As the frequencies of 3 kHz and 6 kHz were tested neither in the hearing aid clinic nor on domiciliary visits, these frequencies were not utilized in mean values for the other two groups. The mean hearing loss (HL) in the domiciliary visit subjects before statistical analysis is the highest for both the sexes, men (at 57.9 dB HL) having a greater loss than women (at 57 dB HL). The least affected population seems to be hospital in-patients for both sexes. The hospitalized women (at 47.4 dB HL) present a greater loss then men (at 45.8 dB HL). Patients seen in the hearing aid clinic appear to be more affected than the population in the Residential Homes.

Statistical Analysis of the Results.   The results were analysed by the method of analysis of covariance (SPSS X User's Guide 1986). There were found to be no significant differences in hearing losses between the two sexes, and no relationship in ordering for either sex among the four groups. There were significant differences present between the four groups. When the hearing losses for the right ear were considered alone, differences between the four groups were found to be significant. The left ear losses revealed differences with even greater significance. After treating the figures for age effect those in the domiciliary visits were confirmed as showing the greatest hearing loss, the

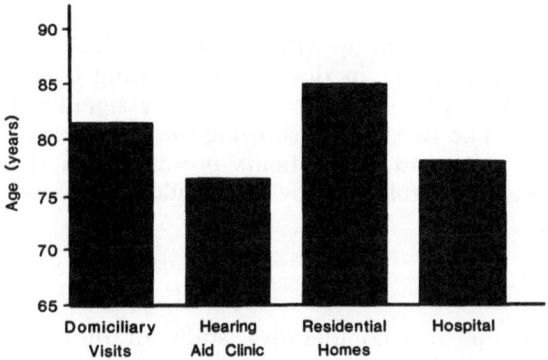

Fig. 2.8. Age distribution for four groups studied in different environmental conditions: domiciliary visits ($n=15$), hearing aid clinic ($n=226$), residential homes ($n=106$) and hospital ($n=76$).

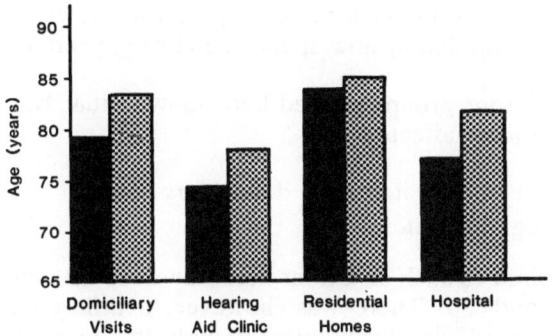

Fig. 2.9. Ages and proportion of sexes among subjects studied in the four groups in Fig. 2.8. (■) men; (▧) women.

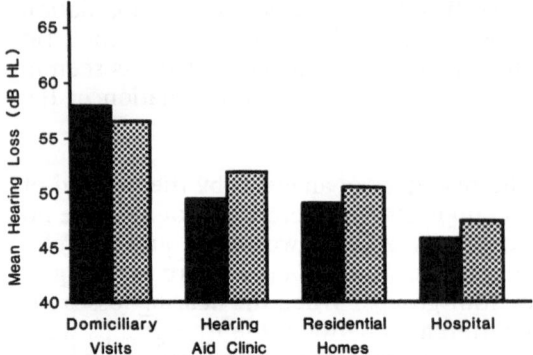

Fig. 2.10. Hearing losses in four groups studied under different environmental conditions. The averages of the means in the right and left ears for the frequencies 250 Hz, 500 Hz and 1, 2, 4 and 8 kHz are shown. (■) men; (▨) women.

lowest value being in the hospital in-patients group with hearing aid clinic and residential homes in an intermediate position, in that order. Hearing in the right and the left ears taken together once more revealed highly significant differences between the four groups. The two groups showing the most severe losses were again the domiciliary visits and the patients presenting in the hearing aid clinic in the descending order followed by the residential homes and the hospital in-patients.

Discussion

That the oldest group of subjects, in this comparative study of different environmental conditions, is represented by the residential homes is not surprising. It has been documented that in the United Kingdom and other developed countries the numbers of people over 80 years of age are growing more rapidly than ever before and increases in the number of very old people have been even greater than that predicted because of falls in mortality rates at advanced ages. This has been noticed particularly over the past decade and a 46% increase in the size of the age group over 85 was projected for 1990 by Grundy (1987). Many of this huge population now spend the closing years of their lives in homes for the elderly.

Analysis of hearing levels of the four groups studied here showed that two groups of the elderly people were most affected:

1. People living in their own homes and visited on a domiciliary visit.
2. People presenting in the hearing aid clinic.

Both groups were hearing-impaired to such an extent that they were actively seeking help for their hearing handicap. There was, however, a difference between them. The former were not sufficiently ambulant to go to the hospital, but not so ill as to be admitted and so had to be examined at home. The reason for their being homebound included blindness, limb immobility or not having an accompanying person to help them with transport. They actively sought social communication and were not worried by the thought of wearing a hearing aid.

The group presenting themselves at the hearing aid clinic were also complaining of deafness and seeking help for this. From this group 97.3% admitted that they had communication problems. Seventy per cent had never had a hearing aid previously. This group was comparable to the previous group, but were mobile and capable of attending the Department of Auditory Rehabilitation.

The group of the elderly living in residential homes represented a sample of people who could not look after themselves, but professional care was not needed on a daily basis. They did not seek any rehabilitative measures for their deafness. In this group 45%, however, did admit to having communication problems, and 25% were hearing aid users.

The least affected group were the in-patients of the hospital. This group, who were receiving the highest intensity of medical care of all four groups, were also the most unwilling to co-operate in the tests although 50% admitted they did have communication problems when questioned. Only 12% were hearing aid users.

Our audiometric studies would, therefore, suggest that the important factor in producing more severe hearing losses in a population of the elderly is their own self-selection, because of greater problems in audition; general medical illness does not seem to play a part. Indeed the generally well subjects are more deaf than the sick ones. These findings would seem to validate those reached after analysis of the literature in Section 1.

It might have been useful to have a further group of elderly people for audiometric analysis, recruiting them from clubs for the elderly, churches, bingo halls or even pubs, so as to compare those who do not require medical care and do not seek any help on account of their hearing problems with those already studied. If the conclusions we reached are valid then the mean hearing losses in such a medically fit group not seeking help for hearing problems would be similar to those in the hospital and residential home groups.

The finding that the elderly living at home who have requested a domiciliary visit for their hearing problems have the worst hearing of the four groups of the elderly population (even though the numbers covered in this study were small) indicates that this must be an enormous problem throughout the whole community. A large financial and professional effort is needed in the auditory rehabilitation of the aged. The group presenting themselves to the hearing aid clinic also showed particularly marked hearing problems and the need for helping these ambulant people must similarly be stressed.

References

Corso JF (1963) Age and sex difference in pure-tone thresholds. Survey of hearing levels from 18 to 65 years. Arch Otolaryngol 77: 385–405

Glorig A (1957) Some medical implications of the 1954 Wisconsin State Fair Hearing Survey. Trans Am Acad Ophthal Otol 61: 160–171

Glorig A, Davis H (1961) Age, noise and hearing loss. Ann Otol 70: 556–571

Glorig, A Nixon J (1962) Hearing loss as a function of age. Laryngoscope 72: 1596–1610

Goetzinger CP, Proud GO, Dirks, D, Embrey J (1961) A study of hearing in advanced age. Arch Otolaryngol 73: 662–674

Grundy E (1987) Community care for the elderly 1976–1984. Br Med J 294: 626–629

Hinchcliffe R (1959) The threshold of hearing of a random sample of rural population. Acta Otolaryngol (Stockh) 50: 411–422

ISO (1984) Acoustics – threshold of hearing by air conduction as a function of age and sex for otologically normal persons. ISO 7029. International Organization for Standardization, Geneva

Jatho K, Heck K (1959) Schwellenaudiometrische Untersuchungen uber die Progredienz und Charakteristik der Alterschwerhoerigkeit in der verschiedenen Lebensabschnitten (zugleich ein Beitrag zur Pathogenese der Presbyakusis). Z Laryngol Rhinol 38: 72–88

Kell RL, Pearson, JCC, Taylor W (1970) Hearing thresholds of an island population in North Scotland. Int Audiol 9; 334–349

Maurer JF, Rupp RR (1979) Hearing and aging. Tactics for intervention. Grune and Stratton, New York

Milne JS (1977) A longitudinal study of hearing loss in older people. Br J Audiol 11: 7–18

Milne JS, Lauder IJ (1975) Pure tone audiometry in older people. Br J Audiol 9: 50–58

Milton JS, Tsokos JO (1983) Statistical methods in the biological and health sciences, International Students' Edn. McGraw-Hill, Maidenhead

Møller MB (1981) Hearing in 70 and 75 year old people; results from a cross sectional and longitudinal population study. Am J Otolaryngol 2: 22–29

Robinson DW (1988) Threshold of hearing as a function of age and sex for the typical unscreened population. Br J Audiol 22: 5–20

Rosen S, Plester D, El-Mofty A, Rosen HV (1964) Relation of hearing loss to cardiovascular disease. Trans Am Acad Ophthal Otol 68: 433–444

Sataloff J, Manduke H (1957) Presbycusis. Arch Otolaryngol 66: 271–274

Spoor A (1967) Presbyacusis values in relation to noise induced hearing loss. Int Audiol 6: 48–57

SPSS X User's Guide (1986) 2nd edn. McGraw-Hill, Maidenhead

Taylor W, Pearson J, Mayir A (1967) Hearing thresholds of a non-noise exposed population in Dundee. Br J Industr Med 24: 114–122

Electrophysiology

In Collaboration with S.M. Mason

In many elderly patients pure tone audiometry has proved to be unreliable in hearing assessment. In Section 2, for instance it was shown that only 159 patients in a large geriatric unit out of a total of 215 initially tested by audiometry gave similar audiometric results on retesting. The other 56 (26%) gave audiometric results which were significantly different because of their inability, at this advanced age, to concentrate and co-operate adequately in the test.

Because of this we set out to attempt to establish the features and the site of disturbance of old age hearing loss without depending on the subject's ability to co-operate. A method highly suitable for giving the objective assessment required was fortunately to hand: the electrophysiological one, and, of its different forms, we used auditory brainstem responses (ABR) and electrocochleography. ABR serve to detect neurological disturbances including small discrete lesions within the substance of the brainstem proper, such as demyelinating processes, punctate haemorrhages or tumours. They provide data, also, from which information might be obtained regarding the character of the auditory system in the elderly. Electrocochleography is available to investigate the features of a disorder in the more peripheral part of the auditory system, intracochlear as well as retrocochlear. Thus by means of electrophysiology a large part of the auditory tract could be surveyed, extending by use of cochlear microphonics from the organ of Corti, through the endocochlear and adjacent retrocochlear region by action potentials, through to the upper auditory tract by ABR. Cortical evoked potentials were not attempted in this work because of their time-consuming nature. Such a test would have been too much to inflict upon elderly patients in addition to the two other electrophysiological studies. Moreover, our aim was not to confirm the validity of the pure tone audiometry, or to detect malingering or central deafness, all of which are well displayed by cortical potentials.

In this section we present a detailed account of the electrophysiological studies which were described in the papers of Soucek et al. (1986) and of Soucek and Mason (1987).

Extratympanic Electrocochleography

For the electrocochleographic studies required in this work we utilized the extratympanic method, so as to avoid the discomfort and possible complications associated with the invasive technique.

An extratympanic method published by Mason et al. (1980) seemed to be ideal for the study of the electrophysiology of the inner ear in frail old people. In this method, action potentials (AP) and cochlear microphonics (CM) could be produced with the same characteristics as those recorded using transtympanic techniques by Eggermont et al. (1974). The isolation of a summating potential was also consistently possible, after elimination of the action potential, by increasing the stimulus rate to adapt the auditory nerve. This non-invasive technique was thus suitable for all aspects of electrocochleography and required no anaesthesia.

In a later work, Singh and Mason (1981) recorded simultaneously extra-tympanic electrocochleograms (ET ECochG) and ABR in clinical cases. This procedure was found to be useful for determining the N_1-V interpeak latency in patients where, due to an inability to identify wave I, brainstem recordings alone were not adequate for the differential diagnosis of deafness and the N_1 was used instead. The method worked out by Mason and his colleagues seemed an appropriate one for analysing the mechanisms of old age hearing loss.

Detailed Investigation of ABR; Preliminary Investigation of ECochG

The first part of the electrophysiological investigation was a preliminary investigation of the elderly auditory system, starting with ABR and then utilizing a method for examination of the cochlea by ET ECochG.

Procedure

Auditory Brainstem Responses

ABR were carried out and analysed using a Medelec Sensor system on 49 patients whose ages ranged from 69 to 94 years with a mean of 73 years.

Eleven adult controls under 36 years of age and with normal hearing were also subjected to the same procedure to obtain data for comparison. Their mean age was 28.8 years.

Non-polarizable electrodes were attached to the vertex and to the mastoid/earlobe region. Stimulation by a click of 0.1 ms with a repetition rate of 10 Hz was used. The signals were filtered through a band pass filter with low frequency cut-off at 3 Hz and high frequency cut-off at 3 kHz. Two time averagings of 1024 sweeps were made for each level of intensity, starting at 100 dBnHL.

It is possible by differing placement of the electrodes to record the waves with their peaks upwards or downwards. The quantitative aspects of the waves are identical in both cases. In this investigation, the electrodes were placed in such a way as to produce peaks pointing downwards. The vertex electrode was plugged into the black socket of the pre-amplifier of the Medelec and the mastoid/earlobe electrode into the red socket.

Extratympanic Electrocochleography

ECochG was carried out by the extratympanic method on eleven cochleas from eleven presbyacutic patients. These were from the same group of patients as were tested audiometrically. Seven cochleas of seven control subjects aged 22 to 45 were also tested by the same method.

In performing extratympanic ECochG by the method of Mason et al. (1980) the active electrode was placed at the 7 o'clock position in the external canal on the right and at 5 o'clock on the left. The impedance was kept below 0–2 k. The ipsilateral earlobe was used as the reference electrode site and the forehead as the ground electrode site. The acoustic stimuli were presented to the patient through the headphone of the Medelec Sensor system and recordings were made on this system. Filter settings on the amplifier were 10 Hz high pass and 3 kHz low pass. The ear canal recordings were usually obtained by the alternating polarity method. In this part of the study, the headphones were not shielded, unlike those of the method of Mason et al. (1980). It was, therefore, not possible to record CM. The sensation level of 10 dBnHL was determined biologically by establishing thresholds for 15 normally hearing young adults.

The APs were recorded starting from an intensity of 100 dBnHL and then lowering it by 10 dB steps.

Results

Auditory Brainstem Responses

Analyses Carried Out.   In the elderly patients wave II was often not recognizable and wave IV was hard to distinguish as it was amalgamated into the complex of waves IV and V. To analyse the ABR material, we therefore measured the latencies for waves I, III and V only. We used the 90 dB hearing level (nHL) for these measurements. We also measured wave V latencies at decreasing stimulus intensities in both sexes firstly taken together and then with elderly women and men taken separately.

We then measured the interpeak intervals I–III and III–V at 90 dBnHL so that any pathological disorder in the higher auditory centres at the appropriate brainstem level could be detected. Consideration was also given to central conduction time (interpeak interval I–V) and this was analysed in age decades in the elderly subjects, to detect whether there was any change in the conduction of nerve impulses in the brainstem due to the ageing process.

Latencies of Waves I, III and V: Increased in Elderly.   The latencies of waves I, III and V were calculated at 90 dBnHL and were found in all the elderly patients to be significantly increased from the controls for both the right and left ears ($p < 0.05$ calculated by Wilcoxon rank sum W test). Let us now consider each of the latencies separately.

Wave I Latency: Longer in Elderly; Relatively Shorter in Elderly Females. Wave I showed a mean latency of 1.74 ms (SD 0.02) for 72 elderly ears (Fig. 3.1). In 17 controls the mean value for wave I latency was 1.58 ms (SD 0.10). The difference was significant ($p < 0.13$ by Wilcoxon rank sum W test). The mean wave I latency in elderly females (1.72 ms–SD 0.02) was shorter than that of elderly males (1.75 ms–SD 0.19). The first wave was often absent in the elderly subjects in both sexes.

Wave III Latency: Right Side of Elderly Longer; Relatively Shorter in Elderly Females.   The mean latency of wave III in the elderly for 69 ears was 3.88 ms (SD 0.21) compared with the control group of 16 ears which gave a latency of 3.72 ms (SD 0.13) (Fig. 3.2). The latency of wave III showed no significant difference on the left (3.89 ms for the elderly females and 3.76 ms for the control group; p 0.16 by Wilcoxon rank sum W 2 tailed test). The right side showed, however, a significant difference ($p < 0.006$). The mean latency for 39 elderly right ears was 3.98 ms (SD 0.52) and the 11 control ears of young people 3.69 ms (SD 0.15). The mean latency showed no difference in a control group for young men, in whom this was 3.67 ms (SD 0.28) from young women in whom the mean latency was 3.69 ms (SD 0.14), but differed in the elderly group, being again longer in men (3.96 ms, SD 0.02) than in women (3.84 ms SD 0.21)

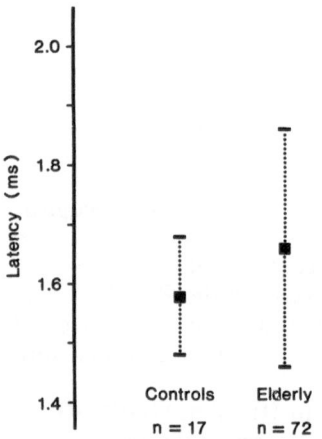

Fig. 3.1.   The mean latency of wave I in the ABR shows an increase in elderly patients compared with controls ($p < 0.03$).

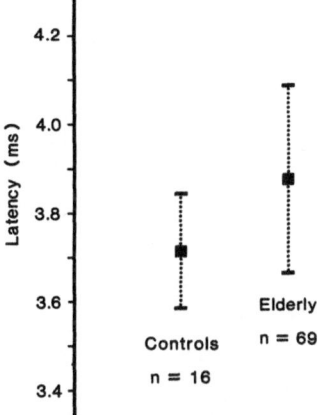

Fig. 3.2. The mean latency of wave III in the ABR shows an increase in elderly patients compared with control ($p < 0.05$).

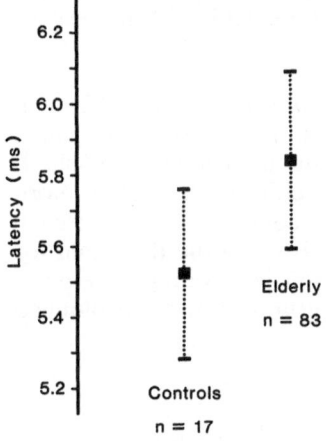

Fig. 3.3. The mean latency of wave V in the ABR shows an increase in elderly patients when compared with controls ($p < 0.05$).

Wave V Latency: Longer in Both Elderly Ears; Relatively Shorter in Elderly Females. The means latency of wave V in 83 elderly people was 5.84 ms (SD 0.25). In controls of 17 young people the mean latency was 5.52 ms (SD 0.24) (Fig. 3.3). A significant difference was found also for the right and left ear taken separately, by the Wilcoxon test ($p < 0.05$), the mean right wave V latency in the elderly being 5.83 ms (SD 0.21), and the mean left wave V latency in the elderly 5.86 ms (SD 0.29). In the young control ears the mean latency for wave V on the right was 5.44 ms (SD 0.24) and on the left 5.63 ms (SD 0.14). The difference in mean latency of wave V was even greater between

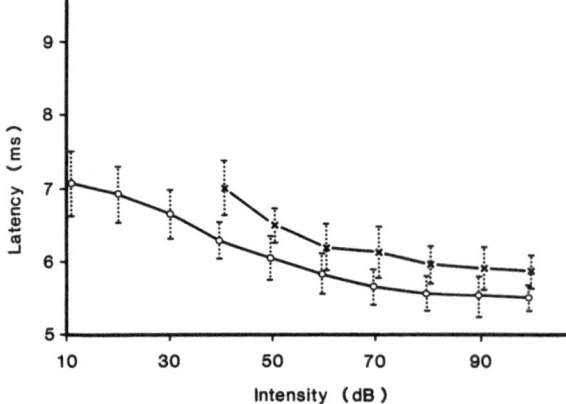

Fig. 3.4. Mean latencies of wave V in the ABR plotted against intensity levels in both ears of 49 elderly patients from geriatric unit (×) and both ears of 11 young controls (O). The elderly show a more prolonged latency for each level of intensity (correlated t-test $p < 0.01$).

elderly men (5.97 ms, SD 0.20) and women (5.75 ms, SD 0.29). The standard deviations were small for both women and men (0.24 ms and 0.20 ms respectively).

Evaluation of Wave V with Decreasing Stimulus Intensity: Longer in Elderly at Each Stimulus Intensity. The mean latencies of wave V were plotted against the intensity levels (Fig. 3.4). It was apparent that as the intensities diminished, the latencies became longer in both groups. The elderly showed a more prolonged latency for each level of intensity than the controls. The difference was significant by a correlated t-test ($p < 0.01$). In this study the threshold for the elderly is shown at 40 dBnHL. A similar situation was found for elderly women and elderly men taken separately. Each group showed prolonged

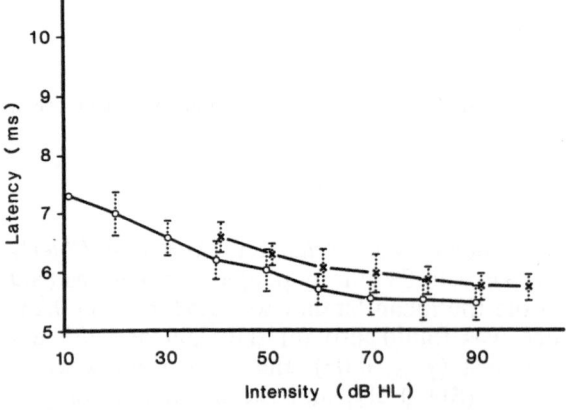

Fig. 3.5. Mean latencies of wave V in the ABR plotted against intensity levels in female patients from geriatric unit (×), compared with young female controls (O). The elderly show prolonged latencies for each level of intensity (t test $p < 0.01$).

latency of wave V with a significant difference from the corresponding group of young people of the same sex ($p < 0.01$ by t-test) (Figs. 3.5 and 3.6).

Interpeak Interval I–III at 90 dBnHL: Identical with Controls; Sexes Equal. The mean interpeak interval I–III showed no significant difference in men, being 2.17 ms (SD 0.26), from females in whom it was 2.11 ms (SD 0.26). Standard deviations were identical for both groups. This interpeak interval (Fig. 3.7) is better compared by considering the control group and elderly subjects with the sexes taken together. An identical figure of 2.13 ms (SD 0.18) for 16 young control ears and 61 elderly ears (SD 0.26) was achieved.

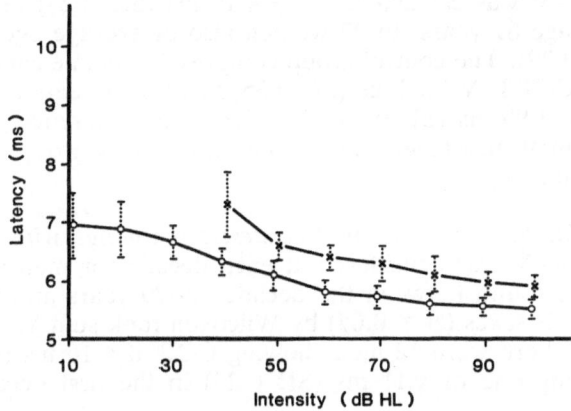

Fig. 3.6. Mean latencies of wave V in the ABR plotted against intensity levels in male patients from geriatric unit (×), compared with young male controls (○). The elderly show prolonged latencies for each level of intensity (t test $p < 0.01$).

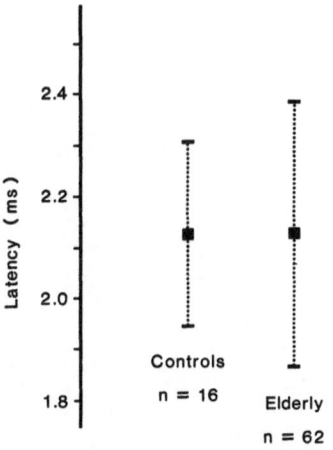

Fig. 3.7. Interpeak interval I–III at 90 dBnHL for elderly patients and controls (both sexes) showing identical means.

Interpeak Interval III–V at 90 dB HL: Right Side Longer in Elderly; Sexes Equal. The mean interpeak interval III–V was calculated in 66 ears of elderly people and found to be 1.93 ms (SD 0.20). The 16 ears of the young control group show the calculated value to be 1.75 ms (SD 0.22) (Fig. 3.8). The Wilcoxon test showed no significant difference between the elderly and young group on the left side ($p > 0.3$). There was a significant difference on the right ($p < 0.05$). When the III–V interval for elderly men and women is considered separately, no significant difference appeared to be present, the elderly group of 41 female ears showing 1.90 ms (SD 0.21) and 25 males showing 1.98 ms (SD 0.18)

Central Conduction Time I–V (CCT I–V) (Interpeak Interval I–V): Identical to Controls. Mean CCT I–V was calculated to be 4.19 ms (SD 0.24) in 26 ears in men of an average age 81 years. In 42 women also of average age 81 years this was 4.03 ms (SD 0.29). The control group comprised 7 female ears of 27 years of age with mean CCT I–V 3.80 ms (SD 0.35) and 9 male ears of 31 years of age with CCT I–V 3.99 ms (SD 0.23). No significant difference was found by Wilcoxon rank sum W test (Fig. 3.9) between the elderly group and the control group as a whole.

Consideration of Central Conduction Time by Decades: No Change with Increasing Age. When CCT I–V was considered at each decade for men and women, no differences were found between the decade 70–79 years and the decade 80–89 years as regards sexes ($p > 0.05$) by Wilcoxon rank sum W test or age groups ($p > 0.05$). There were 12 men showing CCT of 4.16 ms (SD 0.26) and 17 women showing one of 4.11 ms (SD 0.23) in the first decade considered.

The decade of people over 80 years showed CCT values of 4.29 ms (SD 0.21) and 3.99 ms (SD 0.32) for men and women respectively. The decade over 90 in men was calculated to be 4.03 ms (SD 0.22) and in two women 4.03 ms (SD

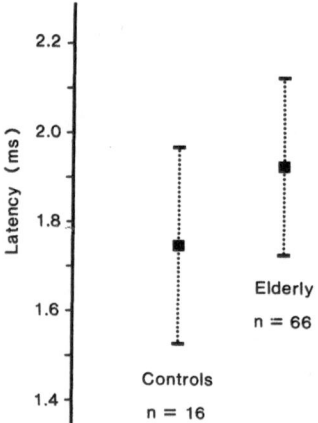

Fig. 3.8. Interpeak interval III–V at 90 dBnHL for elderly patients and controls for both sexes.

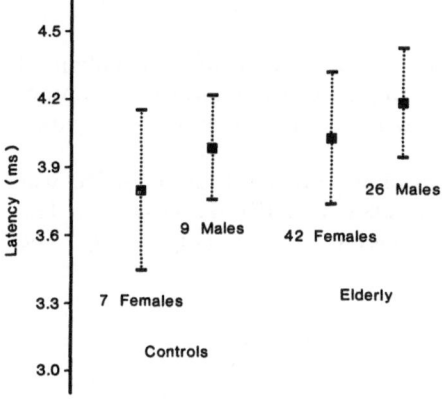

Fig. 3.9. Conduction time I–V at 90 dBnHL for males and females. No significant difference is shown between elderly and controls (Wilcoxon rank sum W test, $p > 0.05$).

0.29). As the number of subjects is so small in this oldest group (two women only) this should be accepted with caution.

Summary of ABR Findings (Table 3.1). In essence, the ABR findings show that the peaks of waves I, III and V were prolonged in the elderly when compared with the young subjects. The interpeak intervals I–III, III–V and I–V, on the other hand, showed similar values in the two groups. This indicates that there is no brainstem lesion in the elderly as the basis for the hearing loss and suggests that this loss emanates from a lesion at the periphery of the auditory system.

Table 3.1. Features of auditory brainstem responses in the elderly

Latency or IPI of wave	Elderly (M + F)	Elderly (F only)	Remarks
I	Prolonged	Shorter	Often absent
III	Prolonged	Shorter	
V	Prolonged	Shorter	Increased with decreasing intensity
IPI I–III	No change	No difference	
IPI III–V	No change	No difference	
CCT I–V	No change	No difference	No change with increasing age

IPI: interpeak interval.
M: males.
F: females.
CCT: central conduction time.

Extratympanic Electrocochleography

Extratympanic electrocochleographic recordings were made of 11 cochleas of elderly people of average age 80 years comprising three men and three women. The control group comprised seven healthy cochleas of average age 29 years.

Range of Latencies.   The range of latencies in the elderly thus recorded was from 1.77 ms (SD 0.14) at 100 dBnHL to 2.25 ms (SD 0.18) at 50 dBnHL. The latter was the threshold of hearing in the elderly subjects (Fig. 3.10). The

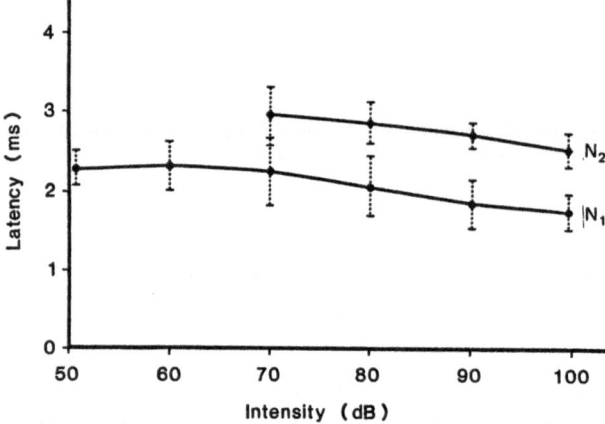

Fig. 3.10. Extratympanic electrocochleography: relation of intensity to average latencies of N_1 and N_2 components of action potentials in 11 cochleas from 11 patients of average age 80 years. The *vertical lines* at each point are the standard deviations of the latencies. Note that the average latencies rise very little below 80 dB. N_2 waves rise slightly in parallel and then cross over at 70 dB.

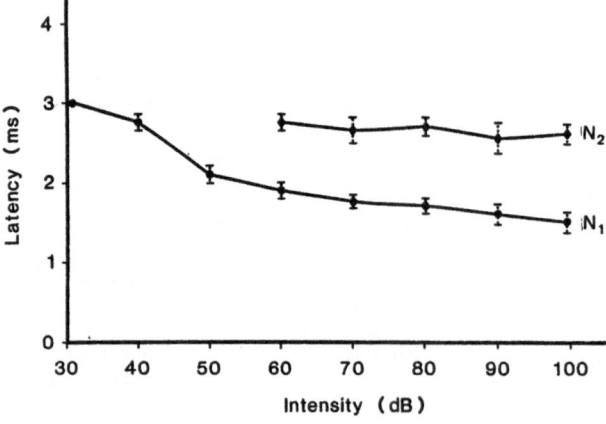

Fig. 3.11. Extratympanic electrocochleography: components of action potential of seven cochleas from seven subjects of average age 29 years. The *vertical lines* at each point are standard deviations of the latencies. The average latencies rise from high intensities. The N_2 crossover takes place at 60 dB.

young cochleas in the controls showed a range of mean latencies from 1.47 ms (SD 0.03) at 100 dBnHL to 3 ms at 30 dBnHL (Fig. 3.11).

Morphology and Latency Function. The AP was ample in the elderly, showing a normal configuration (with summating potential, see below) and its two components N_1 and N_2. The contributions of N_1 and N_2 were followed at the different stimulus intensities in both the elderly group and the young group. The crossover of N_1 to N_2 was recorded at 60 dBHL in the elderly and 50 dBHL in the young group. With increasing intensity level, the rate of decrease of latency of N_1 is greater in the elderly group (0.0177 ± 0.003 ms db^{-1}) than in the control group (0.0113 ± 0.001 ms db^{-1}). However, no significant decrease in rate of latency of N_2 with increasing intensity level was shown for either the elderly or for the control group.

Discussion

Auditory Brainstem Responses

It is possible that the increase in latency of waves I, III and V of the ABR in the elderly is a reflection of the increased sound pressure required to produce the same latency levels as compared with controls. This would be a consequence of the hearing loss that is a feature of all elderly ears (Yamada et al. 1979). Harkins and Lenhard (1980) also suggested that increased peak latency in ABR waves in the healthy elderly may reflect peripheral (cochlear) changes with ageing, not those of central conduction. A similar explanation is given for the changes in the shape of the waves by Stockard et al. (1978), Thomsen et al. (1978) and Yamada et al. (1979). The increased peak latency for all waves of the ABR in the elderly compared to the young have been reported also by other workers (Fujikawa and Weber 1977; Row 1978). Wave V seems to be the wave most sensitive to such peripheral changes and shows a form of recruitment in the transduction process, as well as constituting an index of the pressure wave travelling time (Yamada et al. 1979).

A characteristic feature of cochlear pathology is the presence of recruitment. This is well known as an abnormally rapid growth of the sensation of loudness when the signal intensity is progressively increased above threshold (see Section 2) (Fowler 1936). Formerly it was held that abnormal loudness sensation arose because of a selective loss of activity in lower threshold cochlear fibres (Portmann et al. 1973). It was thought to be the damage to the outer hair cells which produced this deficit (Kiang et al. 1965). This is no longer accepted. Evans (1983) explains recruitment as a selective loss of the more sharply tuned segments of the frequency–threshold curve. This leaves only the higher threshold broad-band segments so that a greater proportion of fibres needs to be recruited for a given increase in suprathreshold tone level. Because of this, loudness grows to a greater intensity than in the normal ear.

The phenomenon of recruitment could conceivably influence the brainstem responses. Indeed, suggestions have been made that ABR could be utilized as an objective means of evaluation of recruitment since the latency/intensity function for wave V parallels the perceptual phenomenon of recruitment very closely (Skinner and Glattke 1977; Coats 1978; Galambos and Hecox 1978;

Yamada et al. 1979; Kavanagh and Beardsley 1979). A steeper latency/ intensity function is obtained in recruiting cochleas. At moderate intensity levels above threshold, the latency of wave V is prolonged compared to that of normal subjects, but at high intensities the latency of wave V is reduced to near normal values as a manifestation of the recruiting hearing loss. The absence of wave V at lower intensity levels in the ABR, evidently the result of the cochlear recruiting effect, is shown in the group of the elderly people as a whole (Fig. 3.4) and also for the group of elderly men taken separately (Fig. 3.6), the loudness/intensity function indicating a steeper slope at higher intensities when compared with the controls. Elderly women, however, do not seem to exhibit a recruiting type of disturbance (Fig. 3.5). The explanation may be that in women there is a different pattern of response of the high frequency cochlear deficit due to a difference in the cochlear transducing mechanism or even that the whole auditory system may not respond to damage by recruitment (Coats and Martin 1977). The shorter latencies of females may also throw some light on this phenomenon. The latency of wave V was not, however, significantly different at each level of intensity when considered for each decade separately for men and women.

The morphological features of the ABR waves were not disarranged. The interpeak intervals I–III were identical in both elderly and young control group. We do not agree that this interval should be used to detect a delayed response due to a peripheral auditory lesion, as Tarkka and Larsen have suggested (1986). Neither should the increased interpeak interval III–V on the right be considered as a sign of a pathological disturbance, as the CCT I–V showed no significant difference on the two sides. The interpeak interval III–V on the right could be an expression of a normal variation in shape of the wave IV–V complex with wave III. As stressed by Chiappa et al. (1973) a number of variations of the ABR tracings have to be accepted as normal and it is likely that this is one of them.

Conduction times I–V showed no significant difference when compared to the control group. This was a consistent finding for both sides in both sexes. When considered for decades, the CCT appeared to be slightly decreased. This was particularly noticeable in women when the decade 90–99 was considered. The number of subjects in this decade was sparse, however, and so this cannot be taken as a conclusive finding. A similar phenomenon was reported by Coats and Martin (1977) and ascribed to high frequency cochlear hearing loss, due to a nonlinear relationship between the higher frequency range (4–8 kHz) and the CCT. No significant differences in the CCT, however, were shown between elderly males and females and between other age groups of the elderly in this study.

Women showed consistently shorter latencies when compared with men. This has been observed previously and anatomical differences have been suggested by several investigators, postulating that females have shorter auditory neural tracts than males (Beagley and Sheldrake 1978; Stockard et al. 1978, 1979; McClelland and McCrea 1979; Jerger and Hall 1980) or that it may be the result of hormonal influences (Fagan and Church 1986). There are, however, no particular variations in ABR latency or amplitude throughout the menstrual cycle, the differences between male and female subjects remaining constant at all times.

Extratympanic Electrocochleography

ET ECochG yielded ample action potentials – a good indication that at least part of the impulses along the spiral neuron were functioning adequately. It has been maintained that an important lesion in old age hearing loss is diminution in numbers of spiral ganglion cells (see Sections 1 and 4). The adequacy of the amplitudes shown by the APs would indicate, however, that loss of spiral neurons is not a functionally important aspect of the hearing impairment in old age. There was always a normal morphology present in the APs of the elderly patients, with the two main components, N_1 and N_2, being recognizable in the recordings.

There was, however, an indication of an abnormality of cochlear function as shown by the relatively small amount of latency/intensity changes and the restricted dynamic range. Such features might, as with the ABR alterations (see above), be indicative of "recruitment". This subject will be considered in greater detail in the description below of the second ECochG study.

Summary

In this preliminary study of auditory brainstem responses in the elderly there were prolonged latencies of wave I, III and V, reflecting a predominantly lower level of activity in the aged inner ear. Cochlear dysfunction seemed to affect the various waves of the ABR differently. The first wave was often missing in the elderly subjects, a feature of a high frequency hearing loss. Women showed consistently shorter latencies than men. Interpeak interval I–III was equal in the elderly compared to controls and so was the central conduction time I–V. These findings suggest that no brainstem deficit is present in old age.

ET ECochG yielded an ample AP of both the N_1 and N_2 components. A specific pattern of input/output function was observed which was interpreted as recruitment. With increasing intensity levels, the rate of decrease of latency of N_1 was greater in the elderly group than in the control group. This was not valid for the N_2 component, however, which showed no significant decrease in rate of latency with increasing intensity level for either the elderly or for the control group. The findings strongly suggest that there is an intracochlear deficit.

Simultaneous Recordings of ABR and ECochG; Detailed Investigation of ECochG Components

In the electrophysiological investigations described above there were deficiencies in the technique which somewhat reduced the value of the findings. The ABR recordings were not made simultaneously with those of the AP. The substitution of the N_1 part of the AP was thus not possible where, as happened sometimes in this study, wave I was not well formed. Such a manoeuvre might have been useful to detect the presence of a brainstem as opposed to a peripheral lesion as the basis of the hearing loss. Because of lack of experience with the technique, moreover, the cochlear microphonic and summating potential were not obtained in the electrocochleogram. These parameters contain information which might be important in our search for sources of cochlear dysfunction. In the following pages a study of electrophysiological findings in elderly and young control subjects using simultaneous recordings of ABR and ECochG was made and the technique was improved so that cochlear microphonics (CM) and summating potentials (SP) could be obtained.

CM is an alternating electrical response produced from the cochlea. It is directly proportional to the displacement of the cochlear partition and therefore reproduces the wave form of the acoustic stimulus. The SP is a direct current response reproducing the wave form of the envelope (or pattern) of the original acoustic signal. SP appears as a shift in the baseline on which the CM are superimposed. Both potentials, CM and SP, are linear and are related to the intensity of the acoustic stimulus. They do not exhibit a "threshold" like AP, which is an "all-or-none" response. There is no evidence to suggest an "all-or-none" response or a refractory period in the activity of the hair cells. The CM and SP correspondingly demonstrate little or no fatigue or adaptation (Glasscock et al. 1981).

Davis et al. (1958) have postulated that the negative SP is mainly generated by inner hair cells (IHC), and that the outer hair cells (OHC) produce both the CM and the positive SP. When sound stimuli are presented, the recorded d.c. potential may either increase or decrease. The duration of the change in amplitude coincides with the duration of the stimulus.

Procedure

Electrophysiological investigations were carried out in this further study on one ear of each of 31 elderly volunteers whose ages ranged from 67 to 97 years (mean age 82 years). They comprised 11 men and 20 women, who were in-patients of a large unit for the elderly, admitted for a variety of illnesses. The control group was composed of studies on a single ear on each of 15 young healthy volunteers of mean age 23 years (range 19–30 years), four men and 11 women. After history taking and clinical examination a subjective pure tone audiogram was performed at eight frequencies across the range from 250 Hz to 8 kHz using an Amplivox 64 audiometer.

ECochG and the ABR were recorded with a Medelec Sensor system. A closed ECochG headphone system was used to present the stimuli monaurally being specially shielded to reduce electrical and magnetic interference. The type of electrode and recording technique for ECochG was again similar to that described by Mason et al. (1980). The Ag/AgCl recording electrode was positioned postero-inferiorly on the canal wall near to the tympanic membrane and was held in place using a conductive bentonite electrode paste. A reference electrode was positioned on the ipsilateral earlobe and, in this study, a guard electrode on the contrateral mastoid. A standard electrode configuration of vertex (active) and ipsilateral earlobe (reference) was used to record the ABR. All electrode contact impedances were kept below 6 kΩ.

The AP waveform of ECochG and the ABR were evoked with an equal number of condensation and rarefaction click stimuli (100 ms duration). The cochlear microphonics were recorded using a 1 kHz tone pip with a one cycle rise/fall time and four cycles on time. Both the click and tone pip were biologically calibrated in decibels above normal hearing threshold (dBnHL) in 10 otologically normal subjects (mean age 23). The peak equivalent SPL for the click and tone pip were 25 and 15 dB respectively.

Filter bandwidths of 5 Hz to 3 kHz for ECochG and 100 Hz to 3 kHz for the ABR were employed. All averaged waveforms consisted of 2048 individual sweeps.

The following components of the ECochG and the ABR were analysed:

1. AP waveforms at a range of stimulus intensities in order to establish the amplitude and latency input/output functions.
2. SP, for evaluation of the proportion of the SP to the total AP waveform.
3. The onset latency and amplitude of the CM at 80 dBnHL.
4. Amplitudes and latencies of waves I, III, V and conduction times I–III, III–V, I–V of the ABR were evaluated at a stimulus intensity of 90 dBnHL. Where wave I was absent in the ABR waveform, the latency of the N_1 potential of the corresponding ECochG trace was used to compute the interpeak latency, since these two functions are generally accepted to have the same origins (Picton and Fitzgerald 1983).

Typical examples of ABR, AP and CM waveforms on which measurements were carried out are shown in Fig. 3.12a and b.

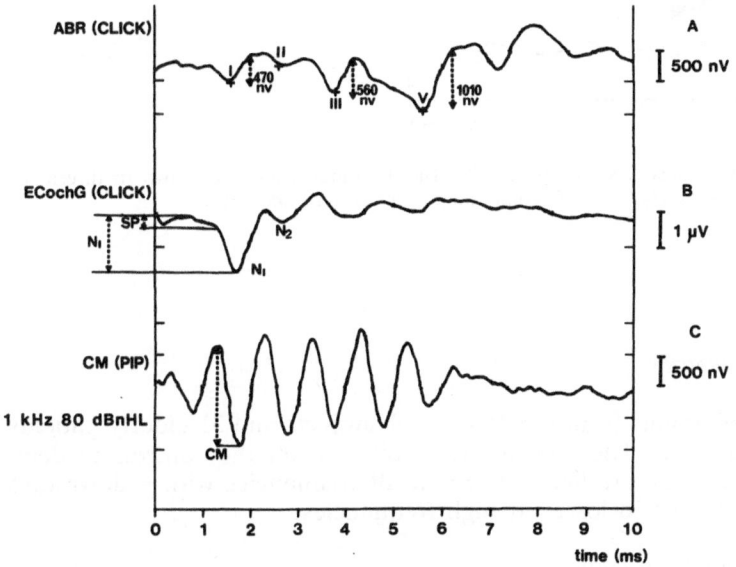

Fig. 3.12.a Normal electrophysiological recordings (female aged 23 years.) Typical examples of waveforms from which measurements were taken. *A*: auditory brainstem response, *B*: action potential, *C*: cochlear microphonics.

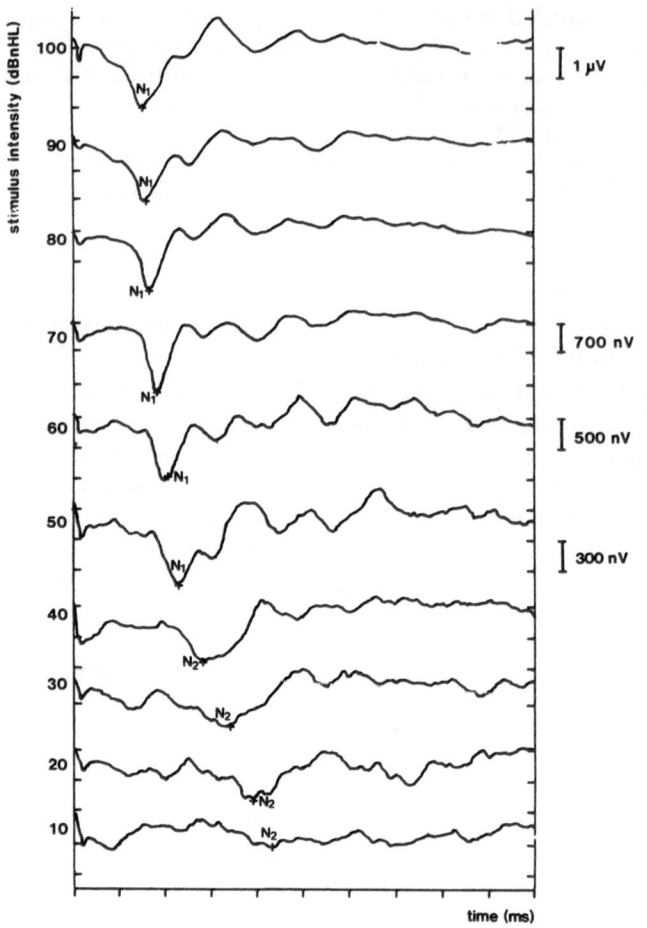

Fig. 3.12.b   Action potentials recorded from 100 dBnHL stimulus intensity down to threshold. Note the crossover from the N_2 to the N_1 component at 50 dBnHL and below.

Results

Pure Tone Audiometry

The averaged audiograms from the 15 control subjects and 31 elderly patients are shown in Fig. 3.13. The elderly group differ from the controls in demonstrating at least moderate hearing loss at all frequencies with a downward sloping curve, and marked losses at high frequencies.

Auditory Brainstem Responses

In the first part of the electrophysiological study described above, amplitudes of ABR were not reported. In the present study ABR amplitudes will be

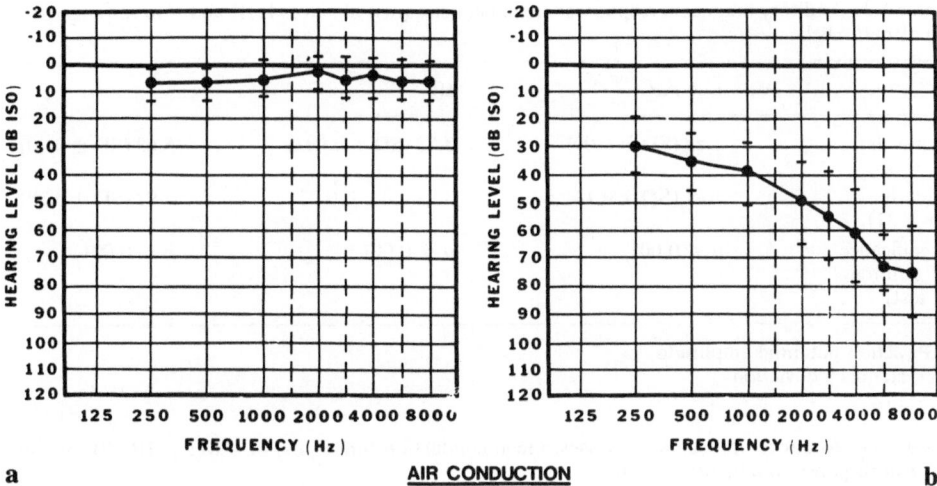

Fig. 3.13. Mean audiograms (±1 SD) for **a** the 15 subjects of the young control group (mean age of 23 years) and **b** the 31 subjects of the elderly group (mean age 81.6 years). There is a marked hearing loss particularly at the higher frequencies in the elderly.

Table 3.2. Auditory brainstem responses: mean amplitudes (nV) of waves I, III and V in response to a 90 dBnHL click

	I or AP	III	V
Controls (n = 15)	469 (SD ± 165)	411 (SD ± 167)	650 (SD ± 166)
Elderly (n = 31)	122 (SD ± 137)[a]	176 (SD ± 147)[b]	423 (SD ± 223)
Significance of difference (t-test)	p < 0.001	p < 0.001	p < 0.001

AP: action potential
SD: standard deviation
[a] Wave I not identifiable in 12 patients.
[b] Wave III not identifiable in one patient.

considered. Latencies of the current study will be briefly presented again for comparison and discussed in relation to the other ABR findings.

There were again no significant delays in brainstem conduction times for the elderly, indicating normal transmission through the brainstem. There was, however, a significant reduction of amplitude for all components of the ABR in the elderly subjects when compared to the controls.

Waves I and III were not identifiable in 12 and 1 elderly patients, respectively. Table 3.2 shows the mean amplitudes of waves I, III, and V in the controls and elderly. All three waves were significantly smaller in the elderly when compared to the controls (t-test: $p < 0.001$ for each wave).

Table 3.3. Auditory brainstem responses: mean latencies (ms) of waves I, III and V in response to a 90 dBnHL click

	I or AP	III	V
Controls (n = 15)	1.53 (SD ± 0.09)	3.73 (SD ± 0.093)	5.60 (SD ± 0.20)
Elderly (n = 31)	1.81 (SD ± 0.14)	3.91 (SD ± 0.23)	5.92 (SD ± 0.20)
Significance of difference (t-test)	$p < 0.001$	$p < 0.001$	$p < 0.001$

AP: action potential amplitude
SD: standard deviation

Table 3.4. Auditory brainstem responses: mean conduction times (ms) of waves I–III, III–V and I–V in response to a 90 dBnHL click

	I–III	III–V	I–V
Controls (n = 15)	2.19 (SD ± 0.13)	1.90 (SD ± 0.20)	4.10 (SD ± 0.22)
Elderly (n = 31)	2.10 (SD ± 0.29)	2.00 (SD ± 0.23)	4.09 (SD ± 0.22)
Significance of difference (t-test)	Not different	Not different	Not different

SD: standard deviation

Table 3.3 shows the mean values of the absolute latencies of waves I, III and V in both the elderly and control groups. The mean latencies in the elderly group were significantly greater than those of the controls (t-test: $p < 0.001$). Conduction times of waves I–III, III–V and I–V were not statistically different in the elderly compared to those of the controls ($p > 0.05$) (Table 3.4).

Action Potentials

In all the elderly patients a well-defined AP was observed showing either N_1 or N_2 components.

AP Amplitudes. The amplitude of the AP was measured from the initial baseline to the dominant N_1 or N_2 component. It was found to be significantly smaller in the elderly compared to the controls (t-test: $p < 0.001$) (Table 3.5). For each subject tested the amplitude at each stimulus intensity was expressed as a percentage of the maximum amplitude of the AP in the same subject; 90 dBnHL for the controls and 100 dBnHL for the elderly. The mean percentage changes at each stimulus intensity were then plotted against the intensity levels. The mean amplitude/intensity curves so obtained indicate an increase in amplitude with stimulus intensity in the elderly which is equivalent to the steeper part of the curve for the controls (Fig. 3.14).

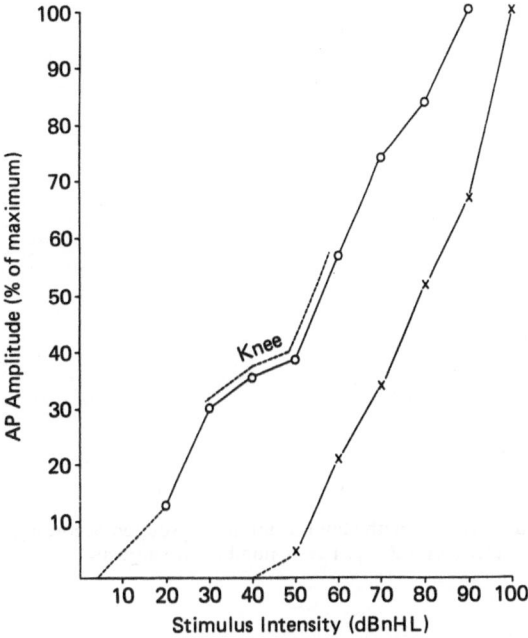

Fig. 3.14. Relationship of stimulus intensity to action potential amplitude expressed as a percentage of maximum AP amplitude (90 dBnHL for the controls and 100 dBnHL for the elderly). There is an increase in amplitude with increase of stimulus intensity. The "knee" in the control group occurs at the crossover from the N_1 to the N_2 component and the steep and shallow curves arise from above and below this crossover level, respectively. (O) Controls, $n = 15$; (×) elderly, $n = 31$.

In the elderly patients the AP thresholds ranged from 90 to 50 dBnHL; three patients at 90 dBnHL, six at 80 dBnHL, four at 70 dBnHL, 11 at 60 dBnHL, and seven at 50 dBnHL. In each of these five threshold groups the AP amplitudes were again expressed at every intensity as a percentage of the amplitude at 100 dBnHL (Fig. 3.15). Across all patients the amplitude/intensity curves were found to exhibit three basic shapes: steep, shallow, or "knee" incorporating both steep and shallow slopes as shown in Fig. 3.14. The steep category comprised 18 patients (58%), the shallow 4 (13%), and the "knee" 9 (29%). The steeper curves suggest a recruiting type of hearing loss, and the shallow ones a lack of recruitment. The "knee" configuration is similar to that of the controls, but is generally present at higher stimulus levels than in the controls.

AP Latencies.   Fig. 3.16 shows the mean latency changes of the AP with stimulus intesity for the elderly and the control groups. As the stimulus intensity is reduced, the latency of the AP is increased in both groups. The group mean latency/intensity function for the elderly is steeper in the range of 60 to 90 dBnHL compared to the controls. The absolute latencies in the elderly at high stimulus levels are consistently longer than in the controls. The latency changes in the elderly were also grouped according to the AP threshold as

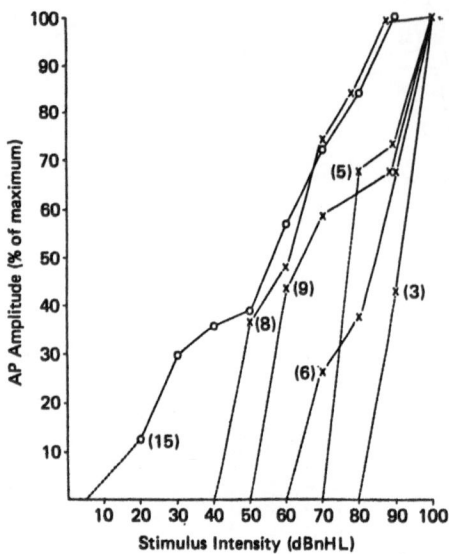

Fig. 3.15. The mean percentage AP amplitude changes with stimulus intensity grouped according to the AP threshold in the elderly. (○) Controls; (×) elderly; (*n*) = number of subjects.

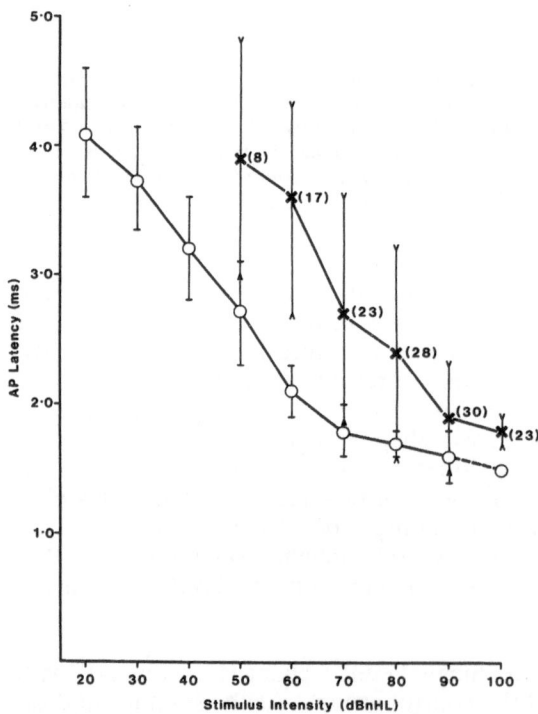

Fig. 3.16. Relationship of action potential latency with stimulus intensity. As the stimulus intensity is reduced, the latency of the action potential is increased in both groups. *Horizontal bars* represent standard deviations for the controls and *arrowheads* represent standard deviations for the elderly. The AP at 100 dBnHL was not recorded in eight elderly subjects due mainly to their intolerance of the stimulus. (○) Controls, *n* = 15; (×) elderly; (*n*) = number of measurements.

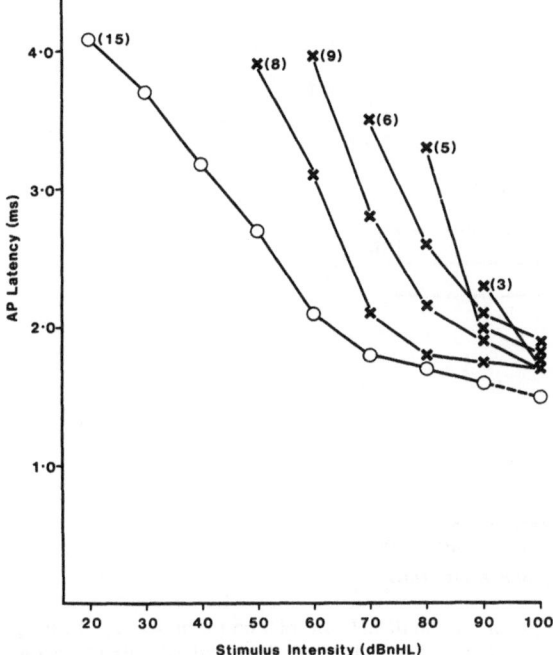

Fig. 3.17. The mean AP latency changes with stimulus intensity grouped according to the AP threshold in the elderly. (○) Controls; (×) elderly (AP threshold groups); (*n*) = number of subjects.

shown in Fig. 3.17. There are five groups of data with mean latencies ranging from 1.74 ms to 1.92 ms at a stimulus level of 100 dBnHL and from 2.28 ms to 3.96 ms at the AP threshold. There is a general increase in threshold AP latency with lower AP threshold.

Fig. 3.18 shows a scattergram of AP latency at 90 dBnHL compared with the average hearing loss across frequencies of 2 kHz, 4 kHz and 8 kHz. It shows that with hearing losses of about 60 dBHL or more there is a higher incidence of long AP latency. A short latency criterion was used as an indication of the presence of recruitment, and patient data were grouped according to whether the AP latency remained less than 3.0 ms. This level was considered optimal after inspection of both control and patient data. There were nine patients who exhibited a short latency function (mean = 2.35 ms).

Summating Potentials

The amplitude of the SP was measured from the initial baseline down to the SP notch of the click-AP waveform. When the SP amplitude was small, identification of the SP notch was occasionally difficult. If this occurred, the recording was carried out again to check for repeatability of the response waveform. The SP amplitude was found to be significantly smaller in the elderly compared to the controls (t-test: $p < 0.001$) (Table 3.5); SP was not identifiable in three

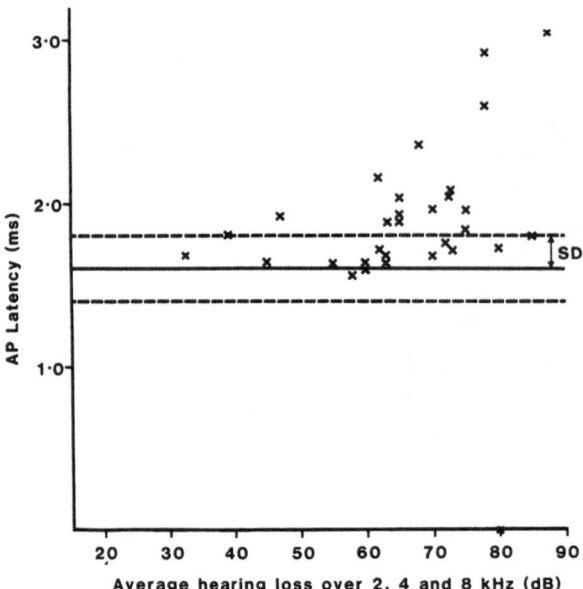

Fig. 3.18. Scattergram of action potential latency at 90 dBnHL in relation to the average hearing loss in dB over 2, 4 and 8 kHz. The *horizontal solid line* represents the mean AP latency of the control group at 90 dBnHL with the *dotted lines* indicating ±1 standard deviation. (×) Elderly.

patients. The relative contribution of the SP component measured as a percentage of the total SP/AP waveform (%SP) was not, however, significantly different in the elderly than in the controls. A mean value of 29% was recorded for each group (SD±15 for the elderly; SD±12 for the controls).

Cochlear Microphonics

Throughout the assessment of the CM, the potential problem of contamination of the averaged waveform with stimulus artefact was taken into consideration. The problem was minimized by achieving low contact impedances for the electrodes, using the specially shielded headphone system and maintaining the stimulus intensity at only 80 dBnHL, rather than at a higher level. An artefact on the waveform could be identified from the time of onset of the "CM" response. An artefact occurred almost instantaneously following the stimulus (< 0.3 ms) whereas a "true" CM waveform had a delay to the onset.

Amplitudes of the CM showed a significant reduction for the elderly compared with the controls; the group mean value for the elderly was 424 nV, (SD±275 nV) and for the control group was 894 nV (SD±650 nV) (t-test: $p < 0.02$) (Table 3.5). The onset latency of the CM, defined as the first deflection of the waveform from the baseline after the stimulus, was also assessed. A group mean value of 0.83 (SD±0.29) was recorded in the controls compared to 0.96 ms (SD±0.44) in the elderly. This difference between the two groups was not significant using a difference of means t-test ($p > 0.05$).

Table 3.5. Extratympanic electrocochleography: mean amplitudes (nV) and latencies (ms) in response to an 80 dBnHL stimulus

	AP amplitude	AP latency	SP amplitude	% SP	CM amplitude	CM onset latency
Controls (n = 15)	588 (SD ± 353)	1.70 (SD ± 0.14)	148 (SD ± 52)	29 (SD ± 12)	894 (SD ± 650)	0.83 (SD ± 0.29)
Elderly[a] (n = 31)	262 (SD ± 156)	2.35 (SD ± 0.73)	75 (SD ± 55)	29 (SD ± 14)	424 (SD ± 275)	0.96 (SD ± 0.45)
Significance of difference (t-test)	p < 0.001	p < 0.001	p < 0.001	Not different (p > 0.05)	p < 0.02	Not different (p > 0.05)

AP: action potential.
SD: standard deviation
CM: cochlear microphonics.
[a] AP, SP and CM not identifiable in 3 patients.

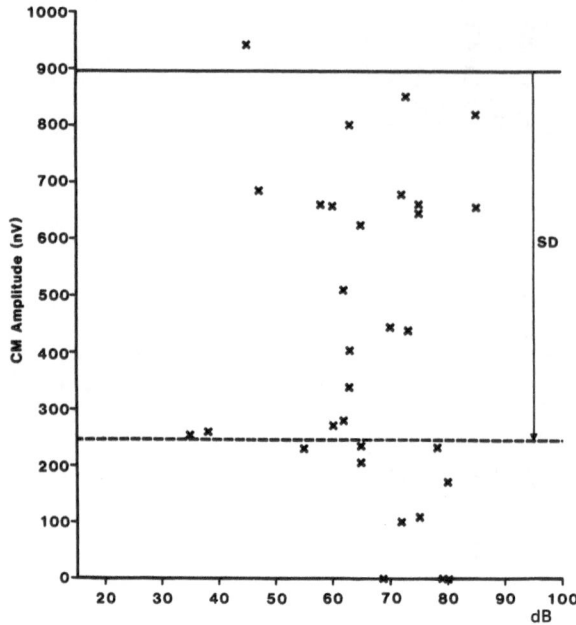

Fig. 3.19. Scattergram of cochlear microphonics amplitude in relation to the average hearing loss in dB over 2, 4 and 8 kHz. The *horizontal solid line* represents the mean CM amplitude of the control group at 80 dBnHL with the *dotted line* indicating −1 standard deviation. (×) Elderly.

The CM amplitude was used as a means of assessing the presence of recruitment and for this purpose the elderly group was divided into two subgroups; those having a CM amplitude of less than 240 nV (10 patients, mean = 128 nV) and those with a CM of greater than 240 nV (21 patients, mean = 566 nV). The level of 240 nV was derived from the control data and was the mean CM amplitude minus one standard deviation.

A scattergram of CM amplitudes against the average hearing loss over 2, 4 and 8 kHz in individual patients is shown in Fig. 3.19. There is evidence of two distinct groups of CM amplitudes since, with high levels of hearing loss (> 50 dB), there are some patients with CM amplitudes that are markedly reduced whereas other patients have amplitudes within the range of the controls. This could be the result of the presence of recruitment. Similar findings were observed when the hearing loss was taken over 1, 2 and 4 kHz. The suggestion of recruitment from the CM amplitude, however, does not correlate significantly as a group with the three classifications of AP amplitude/intensity functions, i.e. steep, shallow and "knee" curves. There are 10 individual patients who exhibit recruitment on both the CM amplitude and the AP amplitude/intensity function, of which only three also had a short latency function.

A decrease of individual CM amplitudes was found with increasing age of the subjects. The degree of this change was determined to be highly significant (Fig. 3.20).

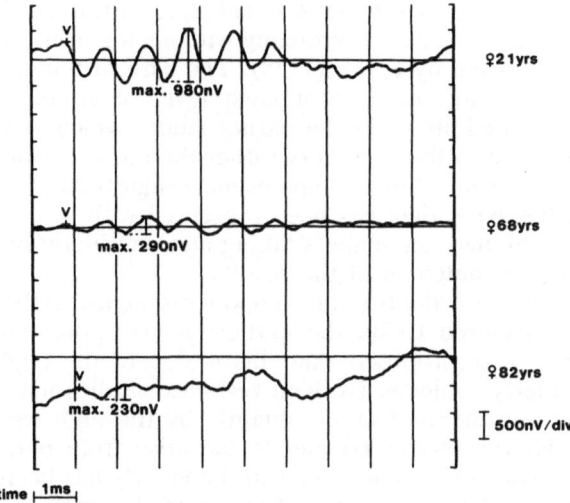

time ⊢1ms⊣

Fig. 3.20. Cochlear microphonic potentials at 80 dBnHL following stimulation pips at 1 kHz, showing decrease of maximum amplitude with increasing age.

Discussion

Auditory Brainstem Responses

The results of ABR in this second study included some features additional to those reported in the first part of this section. For this reason the discussion on ABR will be extended to cover features not previously considered in that earlier part.

Rosenhall et al. (1985) observed prolonged latencies of waves I, III and V with increasing age, but the I–V interpeak interval was the same in all age groups. In this study interpeak intervals I–III, III–V and central conduction time I–V were equal in the elderly and controls. Since the central conduction time is thought to consist of several internuclear segments with axonal propagation and synaptic–dendritic delays (Sohmer 1983) it may be presumed that the patients were free from any neurological brainstem lesions. This finding is not surprising, as we did not include any patients who suffered from disease which could have influenced neural structures and thus caused abnormal transmission. The findings differ from observations of abnormally long interpeak latencies reported in elderly subjects by Otto and McCandless (1982) who concluded that this disturbance is a part of both a central and peripheral auditory age-related change. The mean age of the elderly people studied by those workers was significantly younger by about 14 years than the elderly patients in the present study. Moreover the selection procedure of their subjects was not very specific so that patients with neurological pathology may have been included; in our study we took care to exclude such subjects.

Although prolonged latencies of waves I, III and V of the ABR were observed in this study, Sohmer (1983) has shown that this does not necessarily indicate neurological disorders. Jerger and Hall (1980) also observed a slight age-related effect in the wave V latency of the ABR in subjects with normal hearing; an increase in latency of about 0.02 ms correlating with ageing from 25 to 55 years. This result is also supported by Rowe (1978). The current finding of a significantly prolonged latency for wave V at each level of stimulus intensity confirms the results described above in the earlier study, which was considered to be due to a hearing loss at the periphery rather than to a central disorder. Again, in both test and control subjects, females have slightly shorter latencies than males. Although the ABR data has been presented with results from male and female subjects combined, an analysis taking into consideration the gender effect did not change the outcome of the results.

In the elderly it was not possible to identify all components in the ABR waveform. Wave I, which is considered to be the earliest neural potential was not observed in 12 subjects. It is proposed that this is due to the high frequency hearing loss in the elderly subjects. Derived response studies have shown that wave I of the ABR is generated predominantly by the high frequency region of the cochlea, whereas waves III and V can arise from both high and low frequency regions. The absence of wave I in the elderly has been observed in other studies (Otto and McCandless 1982; Maurizi et al. 1982). An absence of wave II, which is thought to arise from the proximal parts of the eighth nerve, has also been observed in this study and in earlier reports (as mentioned in the earlier study). The peak of the wave III component of the ABR is probably generated predominantly by the cochlear nucleus (Møller and Janetta 1984). The generation of auditory components recorded with latencies greater than wave III may be supported bilaterally by the crossing over of fibres which occurs at the level of the superior olivary complex. We would tentatively suggest that this may explain the absence of an identifiable response (waves I, II and III) from the uncrossed fibres of the lower part of the auditory pathway in the elderly, whereas the late component (wave V) is usually present.

There was a significant reduction in amplitude and increased latency of all components of the ABR. ECochG shows that this is the result of predominantly lower levels of activity in the cochlea, although a smaller contributing factor is also probably the reduced myelination of the nerve tracts, which is a feature of advanced age. Similar observations were reported by Beagley and Sheldrake (1978) and by Maurizi et al. (1982).

Electrocochleography

Action Potentials. In the evaluation of the AP complex it is important to consider the input/output (I/O) functions: the amplitude/intensity and latency/intensity functions. These are inversely related; a lowering of the stimulus intensity is associated with a reduction in response amplitude but an increase in latency. These response characteristics illustrate specific features of a hearing loss such as loudness recruitment.

In the normally hearing ear the I/O curves for both amplitude/intensity and latency/intensity can be subdivided into two parts: a shallow, lower part (L) and a steep, higher part (H). A dual nerve fibre innervation has been postulated to account for the two regions of the curves (Eggermont and Odenthal 1974). More recently it has been suggested by Evans (1983) that the H section at high stimulus intensities arises from activation of large areas of predominantly basal hair cells, being related to the broadly tuned output of the cochlea; whereas the L section corresponds to lower intensity stimuli activating a more frequency specific area of hair cells, being associated with the finely tuned nerve fibre firing.

It has been proposed, however, that fine tuning in fact originates at the hair cell level rather than the neural level, so that there is no need to postulate any neural sharpening (Russell and Sellick 1977). The loss of sharp tuning, which is observed as an elevation in threshold and broadening of the tip of the tuning curve has recently been linked to damage in the outer hair cells (Neely 1985) and this results in a loss of the L section of the I/O curves. In many elderly cochleas a predominance of outer hair cell loss is observed (see Section 5) and in the present study the I/O function of amplitude/intensity exhibited a complete loss of the shallow L curve in 58% of the elderly, suggesting that the nerve fibre activity relating to this section of the curve is lost. The N_2 component of the normal AP is often used to describe the L section of the curve at intensities lower than 50 dB and this component is absent in many of the elderly patients.

Since the elderly cochlea is damaged in other ways than just outer hair cells it would be unreasonable to expect the electrophysiological recordings to show uniform features in the elderly such as complete "recruitment". There have been attempts to interpret a "recruiting" type of hearing loss from the type of I/O pattern of the AP, in terms of relatively short latencies and steep amplitudes (Portmann et al. 1973; also see above). There are three basic types of curves in the amplitude I/O functions: a steep slope, a shallow slope or a "knee" type configuration. Such a heterogeneity of functioning of sensory units has also been found in observations made on the AP recordings in cases of sensorineural hearing loss (Yoshie 1973) and in a group of subjects over 58 years of age (Bergholtz et al. 1977) in which only about half of the subjects showed steep amplitude/intensity curves.

The group mean latency changes of the AP with high stimulus intensity in the elderly subjects show a more rapid increase in latency when compared to the controls. The absolute latency at high stimulus intensities is longer in the elderly than the controls. When the latency functions in the elderly are grouped according to the AP threshold there is a general increase in the end-point latencies with lower AP thresholds. This behaviour of the AP latency can be explained as an absence of a population of nerve fibres which is responsible for the shorter latencies and which has a different dynamic range (Eggermont and Odenthal 1974). However it is likely that atrophy of the basal part of the cochlea is the explanation for the latency delay: the AP responses arising from more apical regions of the cochlea with poor synchronization of firing of nerve fibres. It has been suggested that nearly all of the well-synchronized short latency APs evoked with intense high frequency transient stimuli are due to hair cell and subsequent neural activity arising in the first turn of the cochlea (Deatherage et al. 1959). Elberling (1974) suggested that the normal transition between the N_1 and N_2 components of the AP reflects a shift of the cochlear activity from the basal part towards more apical regions.

Scrutinizing the latency behaviour it appeared that nine patients fulfilled the criteria suggested for "recruitment" – a shorter latency function. Six of these also exhibited a steep amplitude/intensity function. The elderly subjects on the whole do not show the classic latency behaviour for recruitment and this is probably a result of not only loss of outer hair cells but also general atrophy of the basal turn of the cochlea.

Summating Potentials. A negative SP was found which is in agreement with previous reports on this parameter in sensorineural hearing loss (Mori et al. 1982). The SP amplitude was significantly reduced in the elderly, and its ratio to the total AP amplitude (% SP) was equal in both the controls and the elderly. This provides evidence that the sample was not contaminated by cases of endolymphatic hydrops.

Cochlear Microphonics. The concept has been strengthened over the years that the CM is partly the product of OHCs (Dallos et al. 1972); this is supported by intracellular recordings in guinea-pigs of the receptor potentials from OHCs. The concept is now advanced that OHCs may have a mainly motor role, responding rather as effectors to proprioceptive information from their efferent innervation, with doubts being cast about the efficacy of their afferent synapses (Russell and Cody 1984). The CM would thus seem to be partly a manifestation of this motor activity. An additional source of CM is thought to be the IHCs, with an activity proportional to the velocity of the basilar membrane (Dallos et al. 1972).

A "recruiting" phenomenon in the CM has been suggested due to the presence of a faster rising I/O curve of IHCs (Karlan et al. 1972). Twenty-one of the elderly cochleas exhibited larger CM amplitudes (greater than 240 nV), which is a similar number to the 18 patients who showed steep amplitude I/O functions for the N_1 component; 10 patients showed both these features of recruitment. This behaviour of the CM can be explained by the recruiting type activity of IHCs and thus parallels with the behaviour of the AP.

The latency delay in the CM and the diminution in overall amplitude in the elderly when compared to young controls supports reduced contribution to the CM from the hair cells in the basal coil of the cochlea. This is in keeping with histopathological changes which will be described in Section 5.

A recently discovered product of hair cell activity with quite a different form of presentation and detection to the CM is the acoustic emission described by Kemp (1978). This is an energy response which can be detected in the ear canal at some tens of milliseconds after reception of a sound wave. It is thought to be a product of the outer hair cells in its "rebound" reaction. On the assumption that its presence would indicate the healthy activity of the hair cells and its absence lack of function of the latter, it could prove to be useful confirmation of the damaged organ of Corti in presbyacusis. We did indeed attempt to detect the "Kemp echo" in 12 elderly cochleas and a similar number of young controls with the help of Dr David Kemp himself. Unfortunately the ambient conditions were too noisy and the results obtained were not sufficiently reliable to be included in this study. An examination of acoustic emissions by Bonfils et al. (1988) in subjects between 2 and 88 years of age was more successful. Before 60 years the emissions were present in 100% of subjects. After 60 the incidence fell to 35%. This is in keeping with the findings suggesting OHC damage in elderly cochleas which we stress to be the basic lesion of old age hearing loss in the present work.

Conclusions

These electrophysiological investigations are in line with histopathological studies which will be described in Section 4 and confirm that the primary auditory disturbance in the elderly originates in the cochlea. The reduction in amplitudes of the various components of the electrocochleogram, i.e. AP, SP and CM, reflects the lower level of activity in that organ. The reduction in amplitudes of the ABR can be explained by this lower activity of the end organ. The general increase in AP latency and onset of the CM at high stimulus intensity also suggests a predominantly basal hair cell loss. The presence of features of recruitment in the I/O function of the AP and CM suggests that the loss is predominantly due to OHC damage. All these findings are in keeping with pathological studies in the elderly to be described later which show a predominance of OHC loss with general atrophy of both basal and apical coils of the cochlea.

Summary

The ABR exhibited normal interpeak intervals of waves I–III and I–V in a group of 31 elderly people (mean age 82 years), indicating normal conduction through the brainstem. The amplitudes of the waves I, III and V were significantly reduced and the peaks I, III and V showed a delay in latency, reflecting a predominantly lower level of activity in the elderly cochlea.

ET ECochG yielded significantly reduced amplitudes of the N_1 component of the action potential, summating potential and cochlear microphonics in the elderly compared to the young controls. Heterogeneity of patterns of I/O functions was observed and interpreted as either an absence of recruitment, as partial or as complete recruitment. The findings are in keeping with damage to hair cells.

Stimulus Rate Effects

Introduction

Adaptation is a phenomenon in which the amplitudes of the auditory responses decrease and the latencies increase with greater repetitive speed of the stimulus. We have studied this phenomenon also to find out more about the system of transmission in the auditory pathways of the elderly.

Compound APs evoked by short tone bursts or clicks show a dependence on the interstimulus intervals. When the intervals become shorter adaptation manifests itself by the amplitude decreasing and the latency of the first negative deflection increasing (Peake et at. 1962; Eggermont and Spoor 1973; Prijs 1980). ABR show similar adaptative reactions to increased stimulus rates. These electrical responses and also the SP and CM were used to investigate the possible contrasting effects of adaptation in the cochlear and brainstem auditory pathways in elderly subjects and normally hearing young adults.

Procedure

The AP, SP and CM were recorded using the extratympanic ECochG technique combined with simultaneous ABR measurements described above. The recording electrode for the ECochG was again positioned posteroinferiorly on the deep canal wall close to the tympanic membrane and a reference electrode placed on the ipsilateral earlobe. The ABR was recorded using a standard vertex and ipsilateral earlobe electrode configuration. A guard electrode was positioned on the contralateral earlobe for both ECochG and ABR. A Medelec Sensor EP System interfaced to an Apple II microcomputer was used for data collection and analysis. Wide-band click stimuli (generated by electrical pulses of 100 ms duration) and 1 kHz tone pip stimuli (2 cycle rise/fall time and 2 cycle on time) were presented at high and low suprathreshold intensity levels with rates of 5, 10, 20, 50, 100 and 200 per second. The order of presentation of the different stimuli was randomized for each subject so as to reduce any effects of fatigue. High suprathreshold stimulus level was set at 80 dBnHL in the normally hearing subjects and at either 90 or 100 dBnHL in the elderly. The response waveforms were recorded with a sweep time of 10 ms and signal averaging of 2048 individual responses was carried in conjunction with artefact rejection.

Twelve elderly subjects (mean age 76 years) and eight normally hearing young subjects (mean age 23 years) were investigated. Subjects were excluded if there was any evidence of previous noise exposure or underlying otological and neurological complications. The right ear on each subject was tested since this was the most convenient arrangement regarding placement of the ECochG recording electrode using the operating microscope and the evoked potential equipment, the Medelec Sensor.

Results

Mean pure tone audiograms for each subject group are shown in Fig. 3.21. There was at least a moderate hearing loss at all frequencies in all of the elderly and most had severe losses at high frequencies. The young people used as controls had, in contrast, normal hearing with pure tone thresholds of better than 20 dB at frequencies of 250Hz to 8kHz.

Electrocochleography

The typical effect of stimulus rate on the compound AP is shown in Fig. 3.22. There is the expected reduction in amplitude of the neural component of the response combined with an increase in latency as the interstimulus interval increases. However, the SP component of the waveform shows resilience to the effects of stimulus rate. The amplitudes of the APs show a dependence on the interstimulus interval. When the intervals become shorter the effect of adaptation is that the relative amplitude of SP to AP increases, i.e. the percentage of SP increases with increase in stimulus rate and the latency of the first negative deflection increases. Fig. 3.23 shows the percentage amplitudes of SP/AP waveform at increasing stimulus rates. Although the percentage amplitude of the AP component in the elderly is smaller compared to the young controls,

AVERAGE HEARING LOSS

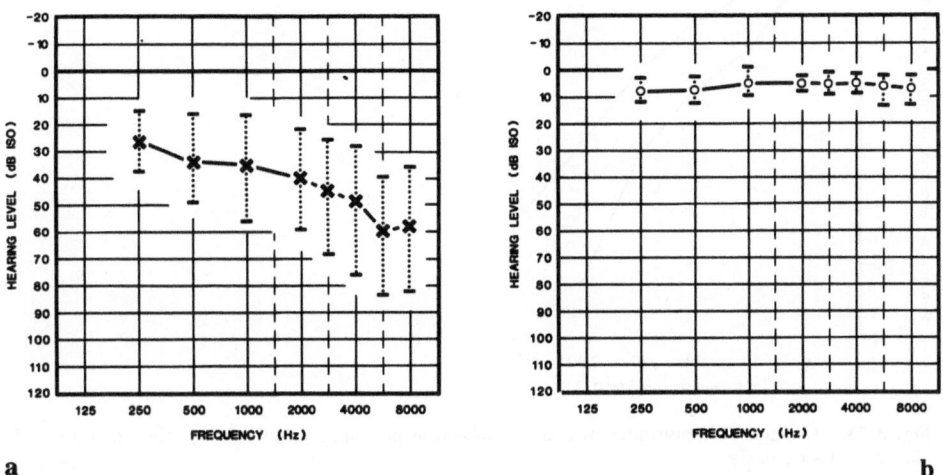

a   b

Fig. 3.21. Mean pure tone audiograms for elderly and control subjects. **a** Elderly, 7 males and 5 females (average age 76.5 years). **b** Controls, 4 males and 4 females (average age 23.4 years).

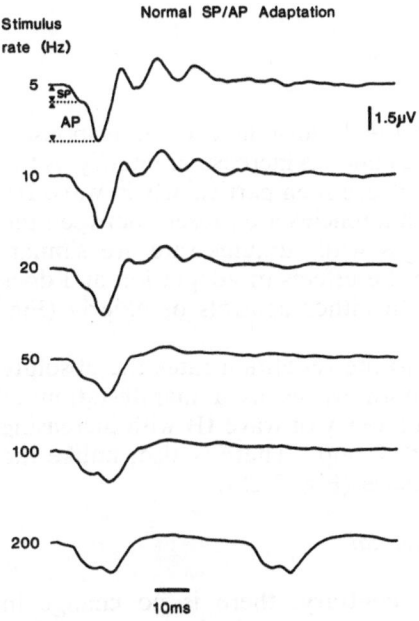

Fig. 3.22. Effect of stimulus rate on the compound action potential in a young subject.

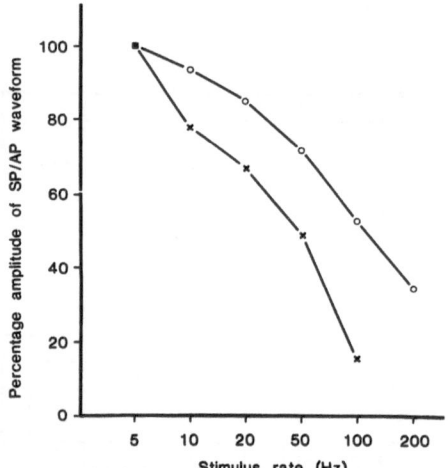

Fig. 3.23. Changes in amplitudes of compound action potentials at increasing stimulus rates. (○) Controls; (×) elderly.

the change in amplitude with stimulus rate is similar in both groups. The curve of adaptation in the elderly is approximately parallel to that of the controls and there is no statistical difference in the adaptation responses of the two groups ($p < 0.05$).

There is also a delay in latency of the AP in the elderly compared to the young controls, but the change in latency with stimulus rate is again similar in both subject groups.

Auditory Brainstem Evoked Responses

A similar adaptation effect is found with the brainstem evoked responses. There is a decrease in amplitude of the waves and an increase in latency as the stimulus repetition rate increases. Such effects are seen particularly in wave III, where differences in absolute amplitudes and latencies are present between the two subject groups (see above), but changes with stimulus rate are similar. Wave V is, however, resilient at all ages to the effects of adaptation and does not show much change with stimulus rate in either controls or elderly (Fig. 3.24).

In the case of ABR latencies in relation to the repetition rates the absolute latencies are delayed in the elderly for both waves as a manifestation of presbyacusis (see above). There is a shift in latency of wave III with increasing speed of stimulus to a similar degree in both groups. There is also, unlike the amplitudes, a similar shift for wave V latencies (Fig. 3.25).

Cochlear Microphonics and Summating Potential

With regard to the CM and SP, on the contrary, there is no change in amplitude with increased stimulus rate. The CM and SP both show a smaller amplitude in the elderly than in the young due to damaged hair cells in the

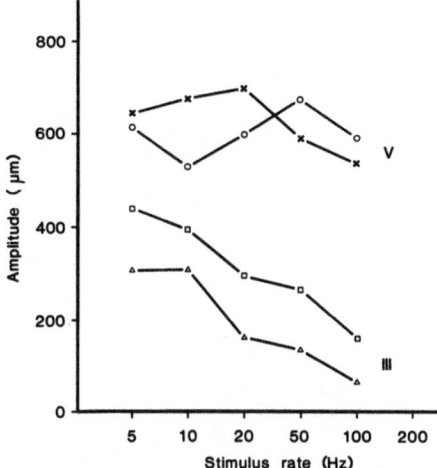

Fig. 3.24. Effect of increase in stimulus rate on amplitudes of waves III and V in controls (□ ○) and elderly (△ ×).

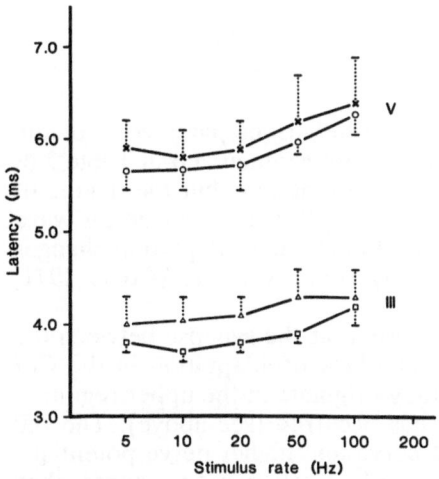

Fig. 3.25. Effect of increase in stimulus rate on latencies of waves III and V in controls (□ ○) and elderly (△ ×).

elderly (see above). In neither group is there a diminution of the amplitude as the stimulus rate is raised (Fig. 3.26).

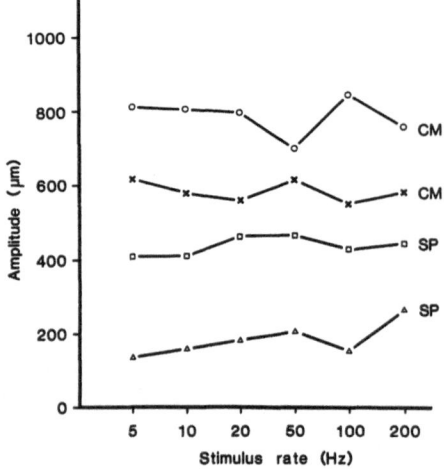

Fig. 3.26. Effect of increased stimulus rate on amplitude of CM and SP in controls (□ ○) and elderly (△ ×).

Discussion

The effects of increasing the stimulus rates on the various parameters of the ECochG and ABR in young controls and elderly are summarized in Table 3.6. APs show adaptation effects in both amplitude and latency, but the degree of change in the elderly is similar to that in the young. The same is true for wave III and for the latency of wave V. CM and SP do not show adaptation changes in either age group. This has been noted by other workers (Coats 1971; Eggermont and Odenthal 1974).

It is thought that adaptation has its major origin at the synapse between the hair cells and the nerve fibre (Prijs 1980). The lack of adaptation in the CM and SP may be explained on the basis that they originate in the upper region of the hair cell itself, i.e. before the synapse has occurred (see above). The AP and ABR originate from the eighth cranial nerve and higher nerve potentials, respectively (see above). Situated central to the hair cell–neuron synapse, they do indeed show adaptation. Elderly people, because of their presbyacutic reduction of hair cells, start off with smaller response amplitudes and longer latencies at the peripheral part of the auditory pathway than younger people, but the effects of adaptation on the AP and ABR are similar in both age groups. This suggests that the synaptic connections to the nerve fibres are functioning normally in the hair cells that survive in the elderly. On these findings the concept of neural presbyacusis, derived as a *morphological* entity from temporal bone histology sections (see Section 1), cannot be considered as a *functional* entity.

Table 3.6. Effects of age on electrophysiological adaptation

	Adaptation	Effect of age on adaptation
Compound action potential:		
Amplitude	+	–
Latency	+	–
Brainstem evoked responses:		
Waves III: amplitude	+	–
latency	+	–
Waves V: amplitude	–	–
latency	+	–
Cochlear microphonics	–	–
Summating potential	–	–

References

Beagley HA, Sheldrake JB (1978) Differences in brainstem response latency with age and sex. Br J Audiol 12: 69–77

Bergholtz LM, Hooper RE, Mehta DC (1977) Electrocochleographic response patterns in a group of patients mainly with presbyacusis. Scand Audiol 6: 3–11
Audiol 6: 3–11

Bonfils P, Bertrand Y, Uzil A (1988) Evoked otoacoustic emissions: normative data and presbycusis. Audiol 27: 27–35

Chiappa K, Gladstone KJ, Young RR (1973) Brainstem auditory evoked responses. Studies of waveform variation in 50 normal human subjects. Arch Neurol 36: 81–87

Coats AC (1971) Depression of click action potential by attenuation, cooling and masking. Acta Otolaryngol (Stockh) (Suppl) 284: 1–9

Coats AC (1978) Human auditory nerve action potentials and brain stem evoked responses. Latency-intensity functions in detection of cochlear and retrocochlear abnormality. Arch Otolaryngol 104: 709–717

Coats AC, Dickey JR (1970) Nonsurgical recording of human auditory nerve action potentials and cochlear microphonics. Ann Otol Rhinol Laryngol 79: 844–852

Coats AC, Martin JL (1977) Human auditory nerve action potentials and brain stem evoked responses: Effects of audiogram shape and lesion location. Arch Otolaryngol 103: 605–622

Dallos P, Billone MC, Durrant JD, Wang C-y, Raynor S (1972) Cochlear inner and outer hair cells: Functional differences. Science 177: 356–358

Davis H, Deatherage BH, Eldredge DH, Smith CA (1958) Summating potentials of the cochlea. Am J Physiol 195: 251–261

Deatherage BH, Eldredge DH, Hallowell D (1959) Latency of action potentials in the cochlea of the guinea pig. J Acoust Soc Am 31: 479–486

Eggermont JJ, Odenthal DW (1974) Action potentials and summating potentials in the normal human cochlea. Acta Otolaryngol (Stockh) (Suppl) 316, 39–61

Eggermont JJ, Spoor A (1973) Cochlear adaptation in guinea pigs. Audiology 12: 193–220

Eggermont JJ, Odenthal DW, Schmidt PH, Spoor A (1974) Electrocochleography: Basic principles and clinical application. Acta Otolaryngol (Stockh) (Suppl) 316

Elberling C (1974) Action potentials along the cochlear partition recorded from the ear canal in man. Scand Audiology 3: 13–19

Evans EF (1983) Pathophysiology of the peripheral hearing mechanism. In: Lutman ME, Haggard MP (eds) Hearing science and hearing disorders. Academic Press, London, Chapter 3

Fagan PL, Church GT (1986) Effect of the menstrual cycle on the auditory brain stem response. Audiology 25: 321–328

Fowler EP (1936) A method for the early detection of otosclerosis. Arch Otolaryngol 24: 731–741

Fujikawa SM, Weber BA (1977) Effects of increased stimulus rate on brainstem electric response (BER) audiometry as a function of age. J Am Audiol Soc 3: 147–150

Galambos R, Hecox KE (1978) Clinical applications of the auditory brain stem response. Otolaryn-gol Clin North Am 11: 709–722

Glasscock ME, Jackson CG, Josey AF (1981) Brain stem electric response audiometry. Brian C. Decker, New York

Harkins SW, Lenhard M (1980) Brainstem auditory evoked potentials in the elderly. In Poon LW (ed) Aging in the 1980's: psychological issues. American Psychological Association, Washington, DC pp 101–114

Jerger J, Hall J (1980) Effects of age and sex on auditory brainstem response. Arch Otolaryngol 106: 387–391

Karlan MS, Tonndorf J, Khanna SM (1972) Dual origin of the cochlear microphonics, inner and outer hair cells. Ann Otol 81: 696–704

Kavanagh KT, Beardsley JV (1979) Brainstem auditory evoked response. Ann Otol Rhinol Laryngol (Suppl) 58, 88: 1–28

Kemp DT (1978) Stimulated acoustic emissions from within the human auditory system. J Acoust Soc Am 64: 1386–1391

Kiang NY, Watenabe T, Thomas EC, Clark LF (1965) Discharge patterns of single fibres in the cat's auditory nerve. MIT Press, Cambridge, Mass

Mason SM, Singh CB, Brown PM (1980) Assessment of non-invasive electrocochleography. J Laryngol Otol 94: 707–718

Maurizi M, Altissimi G, Ottaviani G, Paludetti G, Bambini M (1982) Auditory brainstem re-sponses (ABR) in the aged. Scand Audiol 11: 213–221

McClelland RJ, McCrea RS (1979) Intersubject variability of the auditory-evoked brain stem potentials. Audiology 18: 462–471

Møller AR, Janetta PJF (1984) Neural generators of the brainstem auditory evoked potential. In: Nader RH, Barber C (eds) Evoked potentials II: the 2nd international evoked potential sympo-sium. Cleveland, Ohio, 1982. Butterworth, Boston

Mori N, Saeki K, Matsunaga T, Asai H (1982) Comparison between AP and SP parameters in trans and extratympanic electrocochleography. Audiology 21: 228–241

Neely ST (1985) Micromechanics of the cochlear partition. In: Levin S, Allen JB, Hall JL, Hubbard A, Neely SF, Tubis A (eds) Peripheral auditory mechanisms proceedings, Boston. Springer, Berlin (Lecture notes in biomathematics, vol 64)

Otto WC, McCandless GA (1982) Aging and auditory site of lesion. Ear and Hearing 3: 110–117

Peake WT, Goldstein MH Jr, Kiang NYS (1962) Responses of the auditory nerve to repetitive acoustic stimuli. J Acoust Soc Am 34: 562–570

Picton TW, Fitzgerald PG (1983) A general description of the human auditory evoked potentials. In: Moore EJ (ed) Bases of auditory brain stem evoked responses. Grune and Stratton, New York

Portmann M, Aran JM, Lagourgue P (1973) Testing for "recruitment" by electrocochleography: Preliminary results. Ann Otol Rhinol Laryngol 82: 36–43

Prijs VF (1980) On peripheral auditory adaptation. II. Comparison of electrically and acoustically evoked action potentials in the guinea pig. Acustica 45: 35–47

Rosenhall U, Björkman G, Pedersen K, Kall A (1985) Brainstem auditory evoked potentials in different age groups. Electroenceph Clin Neurophysiol 62: 426–430

Rowe MJ (1978) Normal brain stem auditory evoked responses in young and old adult subjects. Electroenceph Clin Neurophysiol 44: 459–470

Russell IJ, Cody AR (1984) The voltage responses of inner and outer hair cells in the guinea-pig cochlea to low frequency tones. Br J Audiol 18: 253–254

Russell IJ, Sellick PM (1977) Tuning properties of cochlear hair cells. Nature 267: 858–860

Singh CB, Mason SM (1981) Simultaneous recording of extratympanic electrocochleography and brainstem evoked responses in clinical practice. J Laryngol Otol 95: 279–290

Skinner P, Glattke TJ (1977) Electrophysiologic response audiometry: State of the art. J Speech Hearing Dis 42: 179–198

Sohmer HH (1983) Neurologic disorders. In: Moore EJ (ed) Bases of auditory brain-stem evoked responses. Grune and Stratton, New York

Sohmer H, Feinmesser M (1967) Cochlear action potentials recorded from the external ear in man. Ann Otol 76: 427–436

Soucek S, Mason SM (1987) A study of hearing in the elderly using non-invasive electrococh-leography and auditory brain-stem evoked responses. J Otolaryngol 16: 345–353

Soucek S, Michaels L, Frohlich A (1986) Evidence for hair cell degeneration as the primary lesion in hearing loss of the elderly. J Otolaryngol 15: 175–183

Stockard JJ, Stockard JE, Sharbrough FW (1978) Nonpathologic factors influencing brainstem auditory evoked potentials. Am J EEG Technol 18: 177–209

Stockard JE, Stockard JJ, Westmoreland BF, Corfits JL (1979) Brainstem auditory evoked responses. Normal variation as a function of stimulus and subject characteristics. Arch Neurol 36: 823–831

Tarkka IM, Larsen TA (1986) Visual and brainstem auditory evoked potentials in the elderly. In: Proceedings of the 8th Scandinavian congress of gerontology. Jampere, 25–28 May 1986. Exerpta Medica, Amsterdam

Thomsen J, Terkildsen K, Osterhammel P (1978) Auditory brain stem responses in patients with acoustic neuromas. Scand Audiol 7: 179–183

Yamada O, Kodera K, Yagi T (1979) Cochlear processes affecting wave V latency of the auditory evoked brain stem response. A study of patients with sensory hearing loss. Scand Audiol 8: 67–70

Yoshie N (1973) Diagnostic significance of the electrocochleogram in clinical audiometry. Audiology 12: 504–539

Histopathological Changes

In Section 3 the significance attributed to electrophysiological changes in the auditory system of elderly subjects was that of a disturbance of the cochlea, probably in the hair cells. In order to determine the histopathological basis of this change it was necessary to examine the cochlea after death. Findings in this regard will be described in this section. Descriptions of these observations were published by Soucek et al. (1986, 1987).

It would have been most satisfactory if the cases whose cochleas were examined after death had undergone electrophysiological examination of the auditory system during life. Under these circumstances the microscopical analysis would have been carried out with the likelihood of thereby translating the particular functional changes into structural terms. Such a happy matching of physiological and post-mortem studies was not, unfortunately, feasible in this study. Many of the elderly patients that came to necropsy had received audiograms, and a small number, ABR, but none had had ECochG. The reason for this was that fewer electrocochleograms were performed in comparison with the numbers of audiograms and ABR examinations and none of those patients who had electrocochleograms came to autopsy.

In spite of this we would maintain that there *is* validity in the relationship of functional to morphological changes presented in this study. An important feature of the functional observations described in Sections 2 and 3 is that almost all elderly people have severe changes in their hearing and the characteristics of these changes are similar in all elderly subjects. It follows that any morphological abnormalities that are found to be present in all or in the majority of elderly cochleas could be assumed to be related to the functional changes also found in the majority, almost as if one were dealing with both pre- and post-mortem findings in the same individual subjects.

In Section 1 it was pointed out that a frequent problem in the interpretation of published material describing pathological changes in the cochleas of the elderly was the likelihood of autolysis, which would make interpretation of the findings difficult or impossible. For this reason microscopical examination of the hair cells was carried out only on those cases in whom there had been fixation of the inner ear by perfusion through the perilymphatic space. An additional advantage was that in histological sections the functionally most important feature of the hair cells, the stereocilia, was not adequately seen, whereas in surface preparations there structures were well displayed.

For the investigation of a substantial number of cases the drilling method for the examination of the human cochlea is not satisfactory, although it has been the standard method of examination of the organ of Corti (Johnsson and Hawkins 1967). In this method the bony labyrinth is drilled away to reach and sample the membranous labyrinth. The technique is difficult and requires many hours of work for each specimen. Fortunately another method was to hand in our laboratory. This was the microslicing method, which had been devised by Michaels et al. (1983) to improve the system for carrying out temporal bone histological examination. It was now found also to be valuable for opening up the temporal bone as a preliminary to surface specimen examination. Using a special, highly refined cutting device the cochlea was displayed in thin slices, from which the organ of Corti was sampled as a surface preparation. Since the methods of perfusion–fixation and microslicing are fundamental to any work on the morphology of the hair cells of the cochlea they will be outlined here.

Perfusion–Fixation and Slicing of Inner Ear

The technique of perilymphatic perfusion which was described by Iurato et al. (1982) was used. We found that by perfusing inner ears within 10 h of death it was possible to obtain satisfactory results in terms of adequate light microscopical appearances of hair cells. In some cases a longer time can elapse and acceptable appearances still be obtained.

Through an ear speculum the upper posterior part of the tympanic membrane is folded forwards. With a curette any bony overhang is removed to expose the oval and round windows. The incudostapedial joint is divided and the stapes luxated from the oval window. The round window membrane is perforated (Fig. 4.1). A blunted needle of about 1 mm diameter attached to a syringe is used to inject fixative into the oval window. The fixative used for light microscopical investigation is 10% formaldehyde solution. This is preferable to glutaraldehyde in that it penetrates the tissues of the inner ear more completely and so reaches the organ of Corti. Glutaraldehyde should be used if electron microscopy is to be carried out. Fixative is injected through the oval window and observed to emanate through the round window niche region at least 10 times.

The temporal bone is removed at post-mortem in the standard fashion. It is then fixed for at least 4 days. The bone is trimmed and then mounted with molten dental wax on a glass plate, which is itself now mounted on to a metal plinth attached to the inner end of the lever of a special slicing machine (Microslice 2 Precision Annular Saw) (Fig. 4.2). This is a cutting machine with a circular steel blade which is bolted to the machine at 16 points to prevent lateral vibration. Cutting proceeds around a circular inner opening where the blade is tipped with diamond. The cutting edge is lubricated by a continuous

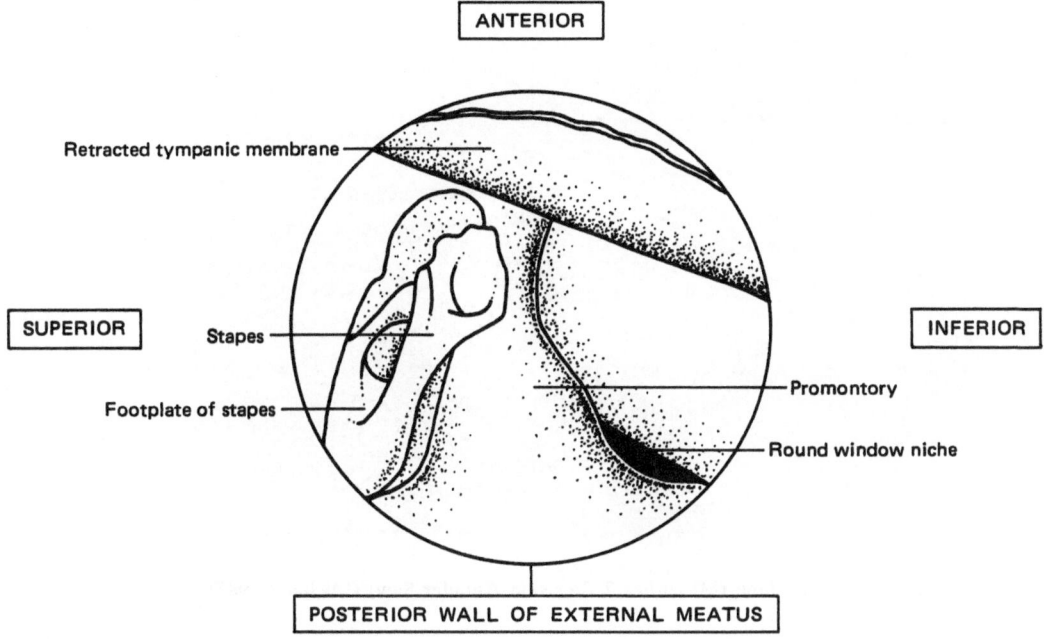

ANTERIOR

Retracted tympanic membrane

SUPERIOR

Stapes

Footplate of stapes

INFERIOR

Promontory

Round window niche

POSTERIOR WALL OF EXTERNAL MEATUS

Fig. 4.1. Diagram of middle ear after retraction of tympanic membrane, showing stapes, promontory and round window niche (Michaels 1987).

jet of cold water. The speed of the rotatory motor may be adjusted. Slicing is carried out by gently lowering the weighted left-hand counterpoised end of the lever so that the specimen rotates up and is applied against the cutting edge. With this system the specimen backs away from the blade when a particularly hard area is encountered, so avoiding excessive mechanical and thermal stresses. The machine is particularly advantageous in preserving the delicate structures of the inner ear with only very slight losses of tissue. Portions of basilar membrane were then removed for surface preparation after carefully excising the tectorial membrane.

Staining of Surface Preparations

It has been the practice up to the present time for surface preparations of the organ of Corti, whether of animal or of human origin, to be examined microscopically after treatment of the specimen with osmic acid solution using the phase contrast technique (Bredberg 1965) or the Nomarski interference technique. In our investigation there was a need to look at hair cells for qualitative changes as well as for their presence or absence. To do this it was necessary to stain these structures. The following staining method was devised for this investigation.

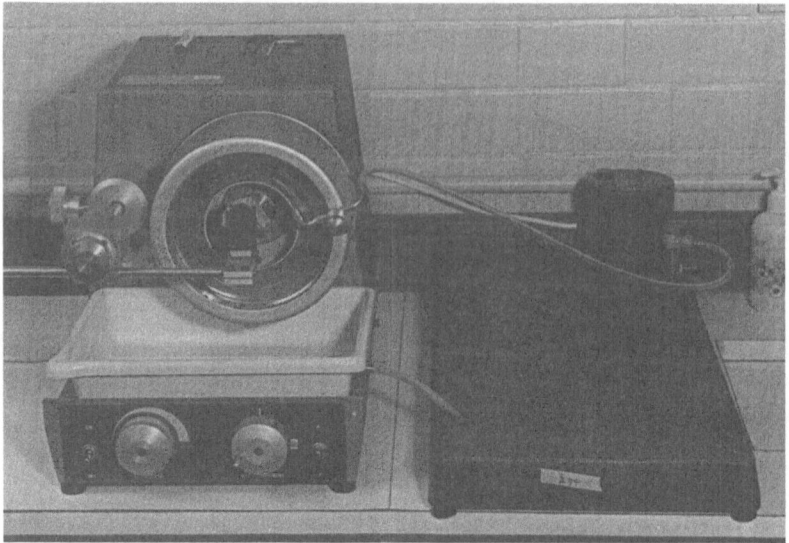

Fig. 4.2. Slicing machine (Microslice 2 Precision Annular Saw) (Michaels 1987).

Microslices of cochlea were exposed to 1% osmium tetroxide for 1 hour and washed four times with distilled water, leaving the specimen in water for 1 min between each change. The slices were then allowed to stand overnight in 70% alcohol to darken. The basilar membrane was removed from the microsliced cochlea in small pieces at positions which were carefully noted. The tectorial membrane was peeled off the specimen. (In some cases it was removed from the basilar membrane in the intact cochlea after microslicing). Each piece of basilar membrane was then exposed to 0.5% Alcian blue solution for 30 s, washed in distilled water and dehydrated in 70% alcohol for 30 s. It was counterstained with eosin-phloxine in 95% alcohol for 3 min. After soaking in absolute alcohol for 30 s the preparation was cleared in terpineol and mounted flat in permanent mounting medium so that the organ of Corti was uppermost. After staining by this method the stereocilia were seen as black structures and the rest of the hair cells brown (Figs 4.3 and 4.5, opposite this page).

Preliminary Survey of Morphological Changes in Elderly Cochleas

Procedure

We examined 25 cochleas from 17 patients at post-mortem by the surface preparation method. Fifteen patients had had pure tone audiometry and five of them, ABR. Five cochleas from three subjects aged 2 weeks, 1 month and 28 years were examined by the same surface preparation method. These served as controls. In each case the perilymphatic space of the cochlea was perfused by

 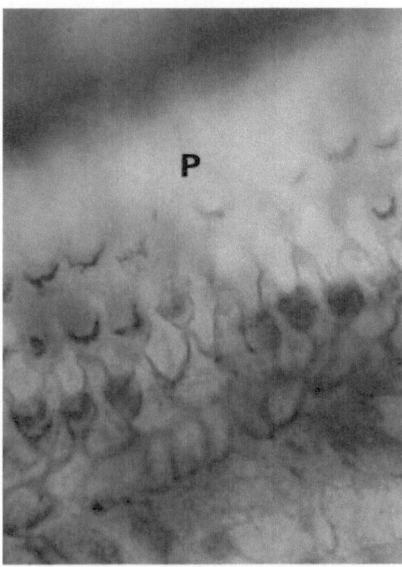

Fig. 4.3. (*top left*)   Surface preparation of basal coil of cochlea in a two-month-old infant showing W-shape formation of stereocilia in outer hair cells, staining black. The first row of outer hair cells is not in focus in this illustration. Osmic acid–Alcian blue.

Fig. 4.4. (*top right*)   Outer hair cell region of basal coil in organ of Corti from an elderly patient. There are gaps among the hair cells of all three rows. *P*: pillar cells. Osmic acid–Alcian blue.

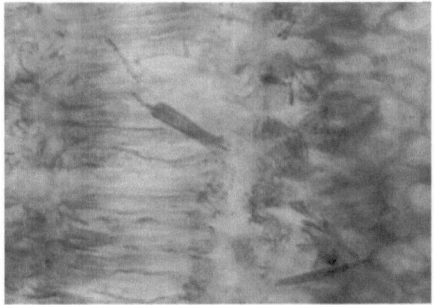 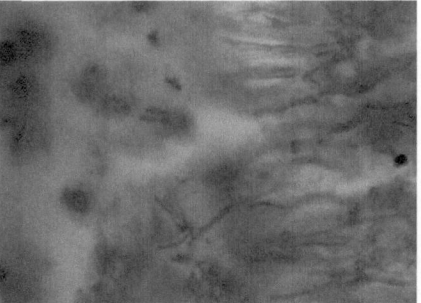

Fig. 4.5. (*bottom left*)   Giant stereociliary change of outer hair cells from middle coil of elderly patient showing marked elongation and thickening of many of the stereocilia. Compare sizes of giant stereocilia with those of normal stereocilia that are present. These are the long structures characteristic of the outer hair cells of the middle and apical coils. Osmic acid–Alcian blue.

Fig. 4.6. (*bottom right*)   Inner hair cells from surface preparation of middle coil of elderly patient showing giant stereociliary change. Osmic acid–Alcian blue.

the technique described above. The temporal bone was then removed and horizontal microslices were prepared at 1 mm thickness by the method described above. After removing Reissner's and the tectorial membranes, samples of the basilar membrane with the organ of Corti, from basal, middle and apical coils were carefully excised and stained as described. Radial nerve fibres were observed in the surface preparation technique with the organ of Corti. Spiral ganglion cells were observed in 15 cochleas from 15 patients by embedding the modiolus in paraffin wax and subjecting it to step sectioning. Stria vascularis and spiral ligament were also processed separately in paraffin wax.

Results

In the specimens prepared by the surface technique two major pathological changes were present:

1. Severe outer hair cell loss.
2. Giant stereociliary degeneration in some of those outer hair cells which survived.

Using such a horizontal slicing method it was not possible easily to relate distances from the end of the basal coil to the position of the hair cells being examined. An accurate method of doing this will be described later in this section. Severe outer hair cell loss was present on inspection of surface preparations of all elderly cochleas (Fig. 4.4, opposite p. 88). and for the purposes of the present study approximate estimates only of the percentages of hair cell loss were made from the samples taken from basal, middle and apical coils in each case. It was found that inner hair cells had rarely lost more than 25% of cells. The first row of outer hair cells showed a greater loss, being deficient in more than 25% in most cases and in a few up to 75% of hair cells. The second row had nearly always lost more than 50% of cells and often up to 75%. The third row showed still greater loss than the second row of outer hair cells, in some cases up to 100% of cells. This pattern of loss was present in apical, middle and basal coils. All three coils appeared to suffer a similar degree of hair cell loss. However, in all cochleas a complete loss of all hair cells, inner and outer rows, was present in a short segment at the lowermost 3 mm approximately of the basal coil. In five cochleas a similar complete loss was also present at the upper tip of the apical coil.

The only other change noted in this study was the presence of enormously lengthened and thickened stereocilia emanating from some surviving hair cells (Fig. 4.5, opposite p. 88). These giant structures were found to measure as much as 60 μm in length. They overlapped many cells in the organ of Corti and sometimes covered the tunnel of Corti. The thickening in some places could be seen to be due to adhesion of hairs to each other as longitudinal lines were identified within enlarged stereocilia (Fig. 4.7). Giant stereocilia were found only in the outer hair cells of middle and apical coils. The outer hair cells of the basal coil never showed this change, although loss of hair cells was just as advanced in this coil. Giant stereociliary degeneration (GSD) was present to a mild degree in the inner hair cell layer of the basal, middle and apical coils (Fig. 4.6).

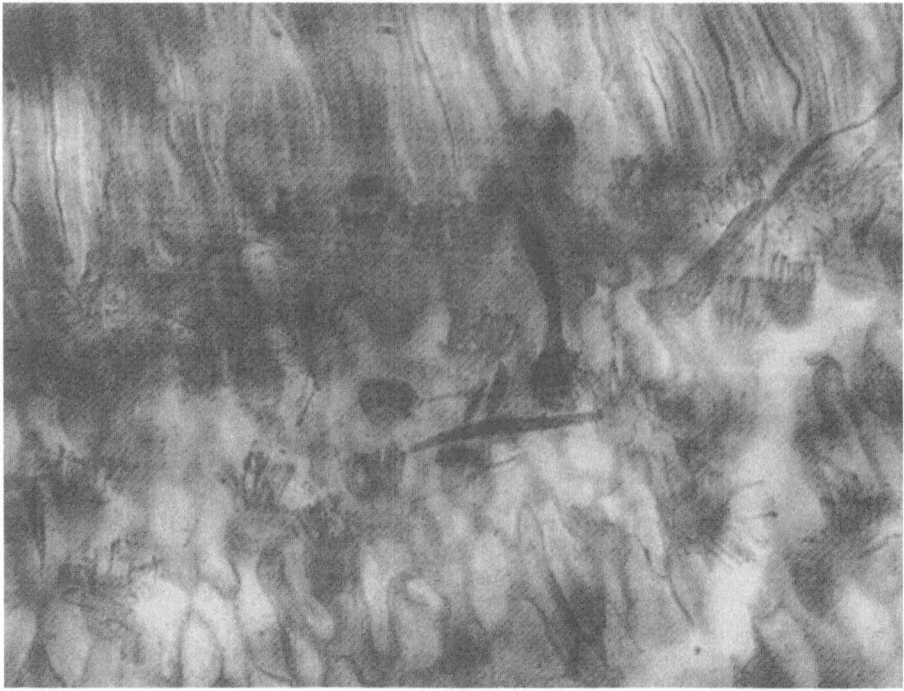

Fig. 4.7. Giant stereociliary change of hair cells from middle coil of elderly patient showing marked elongation and thickening of cilia and vertical lines in some enlarged cilia. Osmic acid–Alcian blue, ×1750.

Severe loss of radial nerve fibres was found only in those situations where there was complete atrophy of all inner and outer hair cells. In the step sections of the normal modiolus cut horizontally, numerous ganglion cells are normally seen in medial and lateral limbs of the basal and middle coils. Few ganglion cells are present in relation to the apical coil, and these may easily by missed since they may occur between mounted sections. Of the 15 cochleas from elderly patients, 10 showed up to between 30 and 50 ganglion cells to be present in each limb of basal and middle coil (Fig. 4.8). This number was seen also in the normal controls. Five elderly cochleas showed a maximum of fewer than 30 ganglion cells in the basal coil limbs, but never fewer than 20 cells; similar numbers were present in the middle coil (Fig. 4.9). These observations would suggest a degree of spiral ganglion cell loss in the latter cases. It was not possible to relate ganglion cell loss to the degree of outer hair cell loss, this latter being severe in each of the elderly patients, whatever the spiral ganglion cell count. No evidence of stria vascularis atrophy was found.

Discussion

A severe outer hair cell loss was shown in all cochlear coils in elderly patients. This confirms the work of Bredberg (1965), who also found outer hair cell loss

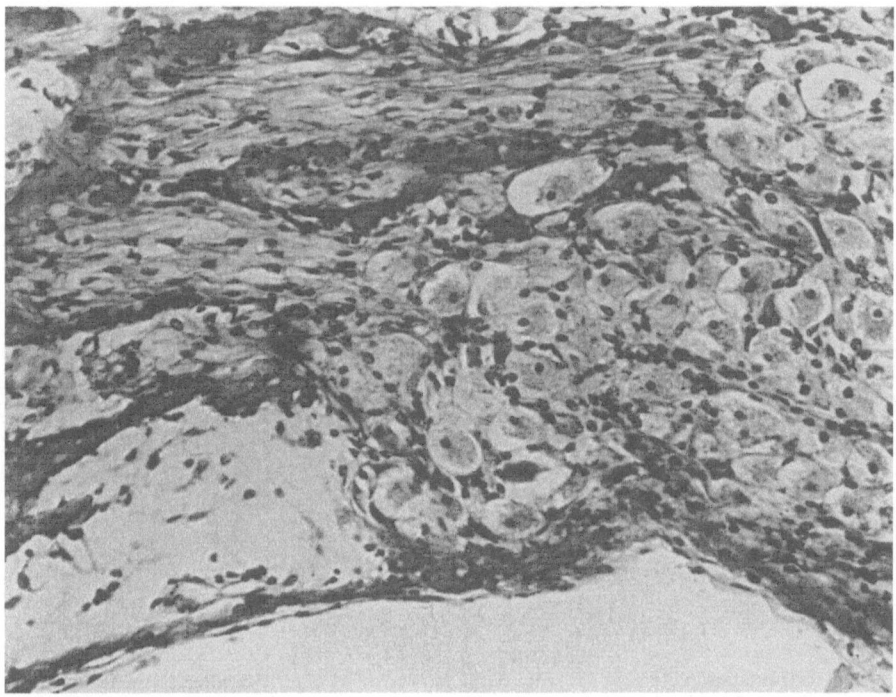

Fig. 4.8. Mid-modiolar section taken from elderly cochlea after basilar membrane has been removed for surface preparation. Note abundant spiral ganglion cells in relation to basal coil. Haematoxylin and eosin, ×150 (Michaels 1987).

to be equally severe in apical, middle and basal coils. The loss of outer hair cells may be the cause of the general hearing disability across all audiometric frequencies. The function of the outer hair cells is not certain. They receive most of the efferent innervation of the hair cell system. Certainly, the outer hair cells are not designed to carry acoustic signal information to the brain. Outer hair cells are capable of a motile response to electrical stimulation (Brownell et al. 1985); inner hair cells do not have this property. The outer hair cell subsystem appears to have the characteristics of a motor system rather than a detection system. The mechanical force generating properties of the outer hair cells together with the resonant tectorial membrane, it is thought, provide the cochlea with an active, second filter which seems to be necessary to explain the high sensitivity and sharp tuning of the cochlea for low level sounds near the threshold of hearing (Neely 1985).

In each cochlea from the geriatric patients there was a short segment at the lower end of the basal coil with complete atrophy of both inner and outer hair cells. This would account for the increased severity of the hearing loss in the higher tones which is a further feature of this age group.

The most significant finding in the present part of this work was the appearance of giant stereociliary degeneration (GSD), often severe, in surviving hair cells. This process includes thickening of groups of stereocilia by adhesion as well as great elongation. Such changes have not been observed by other

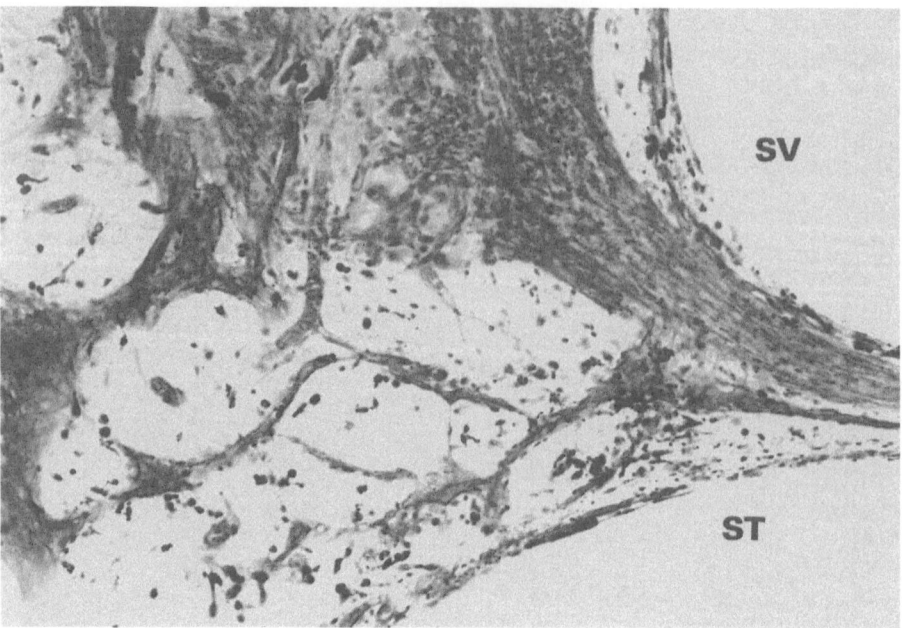

Fig. 4.9. Horizontal modiolar section from basal coil region of elderly cochlea showing some loss of spiral ganglion cells which have been replaced by loose connective tissue. *ST*: scala tympani; *SV*: scala vestibuli. Haematoxylin and eosin, ×120 (Michaels 1987). ·

workers when using the light microscope with phase contrast, or the Nomarski interference microscopy method. Giant stereocilia, it seems, are not visible by these techniques. The staining method which was used here permits these pathological alterations to be clearly observed. The well-defined appearance of the giant structures emanating from non-autolysed hair cells and the fact that they have been observed by other workers using the electron microscope indicates that they are not an artefact of post-mortem autolysis or of histological preparation.

In studies of human temporal bones of age range 80–91 years, in which the organ of Corti was subjected to scanning electron microscopy, giant stereocilia were seen to be growing from the hair cells (Nomura and Kawabata 1978, 1979). They were found to be common in these elderly human cochleas, suggesting that they were a part of the process of ageing. Only part of upper basal and middle turns of outer hair cells were observed. Other parts of the cochlea including the inner hair cells were not studied by this method.

Giant stereocilia have also been seen with the scanning electron microscope in another study of the human cochlea by Wright in people as young as 20 years (Wright 1982). He found them only in the outer hair cells of the apical coil and he observed that their numbers increased with advancing age. A moderate degree of GSD in our material from elderly patients is also present in the inner hair cells of all three coils.

It seems possible that GSD represents a stage in the degeneration of all of the hair cells except the outer cells of the basal coil. In this latter region the stereocilia are normally different in being shorter than in the other two coils. It could be that the changes preliminary to the death of basal coil outer cells do not include a giant stereociliary phase or that this phase is too short-lived in those cells to be identified by histological examination.

The severe outer hair cell loss, as mentioned above would seem to be the basis for the moderate hearing loss across all frequencies in the audiogram as well as the prolonged latencies and reduced amplitudes for waves I, III and V in the ABR and some recruitment features in wave V. This loss would also seem to be responsible for the reduced amplitudes of all three features of the ECochG, the AP, SP and CM and also for the recruitment features seen in the AP and CM. The complete loss of hair cells at the basal end of the cochlea seems clearly related to the marked high tone loss which is the fundamental disturbance of the hearing in old age.

The presence of a severe loss of hair cells in the descending pattern accompanied by GSD in surviving cells in all elderly people suggests that this may be the primary pathological manifestation of presbyacusis. Bredberg (1965) has shown that hair cells begin to disappear from an early age and this, with Wright's observation of giant stereocilia in a 20-year-old (1982), suggests that GSD may be an alteration that is taking place throughout life. Presbyacusis may be a late symptom resulting from this lifelong process; it is not until the passage of many years that there are so few hair cells that hearing would be seriously diminished.

Radial nerve fibres and spiral ganglion cells show a variable degree of atrophy which, as other workers have found (Bredberg 1965; Johnsson and Hawkins 1972) does not correspond with the hair cell loss, unless there is a focal complete atrophy of all rows of hair cells, outer and inner. Experimentally, destruction of all cochlear hair cells by neomycin leads to a slow but incomplete degeneration of cochlear neurons; the retention of a typical, albeit weak, action potential in these animals (Spoendlin and Baumgartner 1977) is a finding which seems to parallel the essentially normal pattern of the action potential tracings in our elderly patients.

Since almost every one of the patients whose cochleas were examined by surface preparation had had audiograms and five had had ABR, it was possible to correlate audiograms and ABR with histopathological changes in specific patients. As mentioned above, the abnormalities were similar in elderly patients for each of the parameters tested (audiometry, ABR, ECochG and hair cell appearances in surface preparations), so that it is also reasonable to relate hair cell morphology to ECochG appearances even though no cochlea was obtained at post-mortem from patients on whom ECochG had been carried out. Correlations are summarized in Table 4.1.

Loss of hair cells accompanied by GSD in animals has been observed by electron microscopy in several conditions. Rabbits subjected to intense sound, for example, can produce both effects, but the giant stereociliary alterations are seen only in the inner hair cells and not at all in the outer hair cells (Engström et al. 1983). In the waltzing hereditary disease of the guinea-pig, the hair cells start to degenerate after birth. The pattern of degeneration is a descending one and giant and fused hairs are observed on both inner and outer hair cells (Ernstson 1971). Although in this guinea-pig disturbance severe changes in the hair cells start and progress in the early years of life, the morphology and topography of the disease process in both waltzing guinea-pigs

Table 4.1. Correlation of post-mortem cochlear changes in elderly patients with audiometry, brainstem evoked responses and extratympanic electrocochleograms

Post-mortem changes	Audiometry	ABR	ECochG
Outer hair cell loss	Moderate hearing loss across all frequencies	Prolonged latencies for I, III and V and some recruitment features in wave V	Reduced amplitudes in AP, SP and CM. Some recruitment features in AP and CM
Atrophy of basal end of cochlea	High tone loss	Reduced amplitudes of waves I, III and V	Prolonged onset latency of CM. Recruitment features of action potential and CM

Table 4.2. Loss of hair cells (+) and giant stereociliary degeneration (GSD) in animals

	Inner	Outer		
		1	2	3
Noise	+ GSD	++	+++	++++
Aminoglycoside ototoxicity	+	++	+++	++++
Hereditary waltzing guinea-pig	+ GSD	++	+++	++++ GSD
Presbyacusis	+ GSD	++	+++	++++ GSD

and in presbyacusis have considerable similarities. However, the changes of presbyacusis in the hair cells differ significantly from noise trauma (Table 4.2). Presbyacusis may thus be the result of a lifelong process which is not related to extrinsic influences.

The observations described above on stained surface preparations of elderly people would suggest that GSD may be an important alteration in hair cells, inner as well as outer, undergoing the processes of ageing. The hair cells of the basal coil do not show it. The stereocilia of these cells are normally short and it is possible that minor degrees of a similar change may be present which are not detectable by the light or even scanning electron microscope. Indeed lesser changes may be common, not culminating in the severe forms of GSD until a terminal phase in the life of the hair cell.

It is certain that the understanding of pathological changes in the hair cell such as GSD will be enhanced by recent discoveries in relation to the structure, chemistry and development of the stereocilia. These have been found to be composed of actin, a filamentous protein present in all cells, but highly concentrated within stereocilia as in the villi of the epithelial cells of the small intestine. The filaments of actin are laid down within the stereocilia in a highly regular fashion (Saunders et al. 1985). The regularity of stereocilial organization seems to extend on a larger scale to the actual dimensions of these structures. In the chick it has been found that the length, number, width and distribution of the stereocilia within the cochlea are predetermined (Tilney and Saunders 1983). It would seem from such work that the precise dimensions of its stereocilia are coded in the very genome of the hair cell nucleus. The alteration of GSD is evidently a profound disturbance in the metabolism of the hair cell.

A concept of the biological processes fundamental to ageing has recently come into prominence. It stresses the inevitable mortality of the somatic cells in contrast to the truly immortal reproductive cells, and indicates that the running down of the metabolic processes of each cell that occurs with ageing is not a genetic process but an intrinsic feature of each cell's activity (Kirkwood 1984). If this is the case then GSC should be regarded as such an ageing aberration of the hair cell, a final common pathway of the ageing cell resulting from faults in the highly complex metabolism required to produce such structures as stereocilia.

Quantitative Analysis of Hair Cell Changes

We describe here a modification by which any particular region of the organ of Corti can be characterized by the ratio of its distance from the basal end to the total length of the cochlea. We quantify, furthermore, some of the pathological changes observed by means of the surface preparation technique, which were mentioned in the first part of this Section. The object of this study was to establish in quantitative terms the pathological changes that occur in the organ of Corti as a result of the ageing process.

Procedure

In each case the cochlea was perfused by injecting formaldehyde solution into the perilymphatic space within 24 h of death by the method described above. In 19 temporal bones from 14 patients over 65 years of age and in four from three infants less than 1 year old, horizontal slices of undecalcified cochlea 1.5 mm thick were, as in the earlier work, cut by the method of Michaels et al. (1983). Samples of basilar membrane were excised from basal, middle and apical coils, and, after removing Reissner's and the tectorial membranes, were stained by the method given above. Counts of inner and outer hair cells were carried out on five microscopic fields of measured length and median values per mm were estimated.

In 13 cochleas from eight patients a method was used in which samples were removed at points which could be easily characterized in relation to their positions in the cochlea. The temporal bone was trimmed to about 15 × 10 mm to include the whole cochlea. It was then X-rayed with the internal auditory meatus downwards on the X-ray plate. The radiograph so obtained was enlarged to 10 × in a photographic enlarger and the configuration of the outer surface of the cochlea coil was drawn on paper (Fig. 4.10).

The cochlea within the temporal bone block was then cut vertically in the plane of the long axis of the petrous bone at 1.5 mm thickness on a microslicing machine. The whole cochlea was usually represented in three slices. With practice, the whole of the cochlear duct through its two-and-a-half coils, from round window to apex, could be identified by moving from one slice to another to follow the continuity of the duct. The basilar membrane was loosened by cutting it away from the bone in the region of the stria on the outer side and through the radial nerves on the inner side. Segments suitable for mounting flat on a slide were then cut out, stained and mounted. The position of each

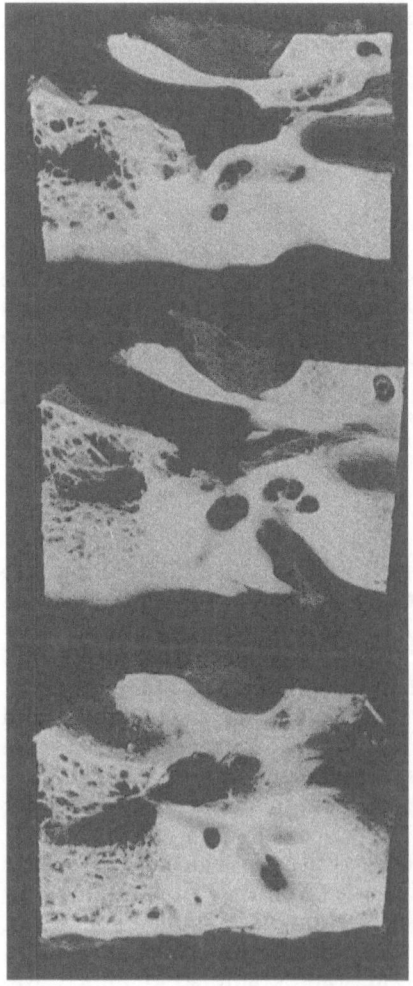

Fig. 4.10. X-ray photographs taken from three microslices of cochlea from one patient. By enlargement to an exact dimension of ×10, the outline of the edge of the cochlea can be drawn to give exact ratios of distance per length for the segments of basilar membrane removed for surface preparation.

segment was marked on to the diagram. The ratio of distance from the base to the total length of the cochlea coils (D/L) of any part of the organ of Corti could thus be obtained by measurements on the diagram.

Results

Findings by Horizontal Slicing Method

Outer Hair Cells. In this part of the investigation, in which samples were taken from apical, middle and basal coils (approximately 29–37 mm, 21–29 mm and 0–21 mm, respectively, from the round window end of the cochlea), the numbers of hair cells per mm were counted separately in the first, second and third rows of the outer hair cells. Findings for rows 1 and 3 in elderly patients are displayed in Fig. 4.11 and 4.12, respectively. In the same figures, counts are depicted for the two rows in four infant cochleas. In most of the elderly cochleas, outer hair cells are fewer than in the cochleas of infants. In a few cases, usually in the apical or middle coils, the levels of first row cells are similar to those in the infants. In most of the elderly patients the basal coil first row counts are lower than the middle or apical ones. In the third rows the counts are in general much lower and all were substantially below those of the infants. Many are near or at zero. The counts of the second rows of outer hair cells were found to be approximately intermediate in value between those of the first and second rows in each case.

Inner Hair Cells. In some of the specimens in which perfusion is delayed for more than about 10 h after death, autolysis of inner hair cells takes place and it

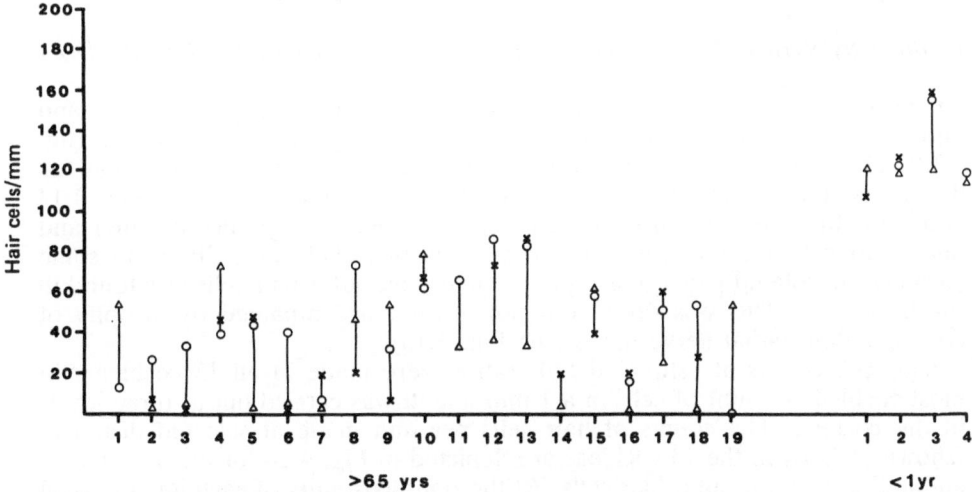

Fig. 4.11. First row of outer hair cells in elderly patients (> 64 years) and infants (< 1 year). Each cochlea is represented by a numbered line on which apical (×), middle (○) and basal (△) coil outer hair cell counts per mm are shown (Soucek et al. 1987).

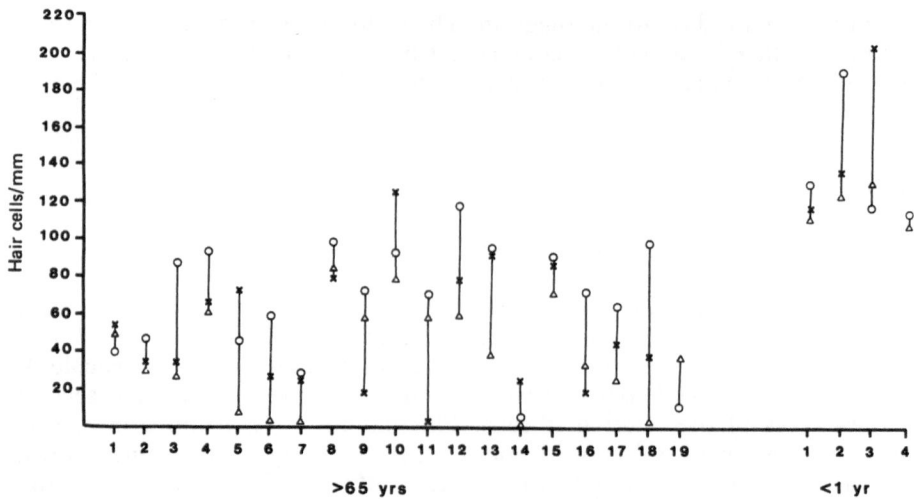

Fig. 4.12. Third row of outer hair cells from the same elderly patients and infants as in Fig. 4.12. Again, each case is represented by a numbered line in which apical (×), middle (○) and basal (△) coil hair cell counts per mm are indicated (Soucek et al. 1987).

is impossible to perform a satisfactory count. Counts were, however, achieved in the majority of cases and showed levels of inner hair cells which were moderately below those of the infants in the apical and middle coils. In the basal coil, the inner hair cells were markedly reduced in numbers and values were near or at zero in some cochleas which were, however, well preserved.

Atrophic Areas. In all cochleas, every hair cell, both inner and outer, was missing in a segment in the lowermost region of the basal coil.

Findings by Vertical Slicing Method with Estimations of Distance/Length (D/L)

At the lower end of the basal coil an area of complete atrophy of inner and outer hair cells and also of radial nerve fibres was present in each case (fig. 4.13). This atrophic area varied in length from 2 to 14 mm and is shown as the flat line at the extreme basal end of the distance/length diagram in Fig. 4.14 and 4.15. In some cases the adjacent region showed an absence of inner and outer hair cells for up to 9 mm, but survival of radial nerve fibres. In some cochleas an isolated patch of atrophy of inner and outer hair cells was found in the basal coil. This was often, but not always, accompanied by atrophy of corresponding radial nerve fibres (see Fig. 4.13).

Hair cell counts at estimated D/L ratios were made in all 13 cochleas. In most cochleas a count of cells in a 1 mm length was carried out at three levels in the cochlea. The counts of hair cells per mm made at selected distances (shown as D/L) in the 13 cochleas are depicted in Fig. 4.14 for inner hair cells and in Fig. 4.15 for outer hair cells. At the right extremity of each line the total length of basal coil complete hair cell atrophy (with and without radial nerve atrophy) is shown as a horizontal line at zero hair cells per mm corresponding to its D/L. To the left of that, the hair cell count levels are joined to form a

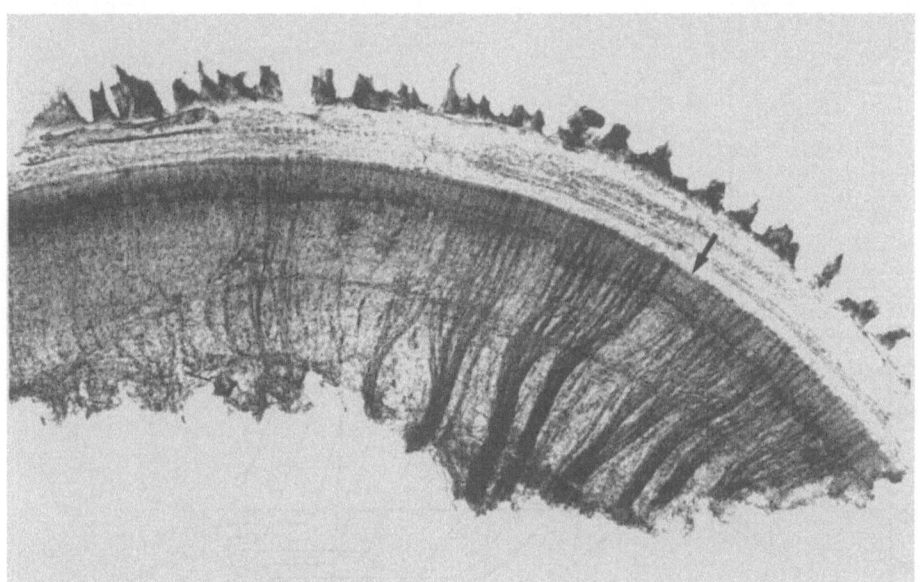

Fig. 4.13. Surface preparation of commencement of basal coil in an 87-year-old man. The hair cells and radial nerve fibres are completely atrophic on one side. There is a small patch where hair cells are absent in the position marked by the *arrow*. The radial nerve fibres in this region also show a small zone of clearing. The serrated edge is produced by adherent stria. Osmic acid–Alcian blue, × 50 (Soucek et al. 1987).

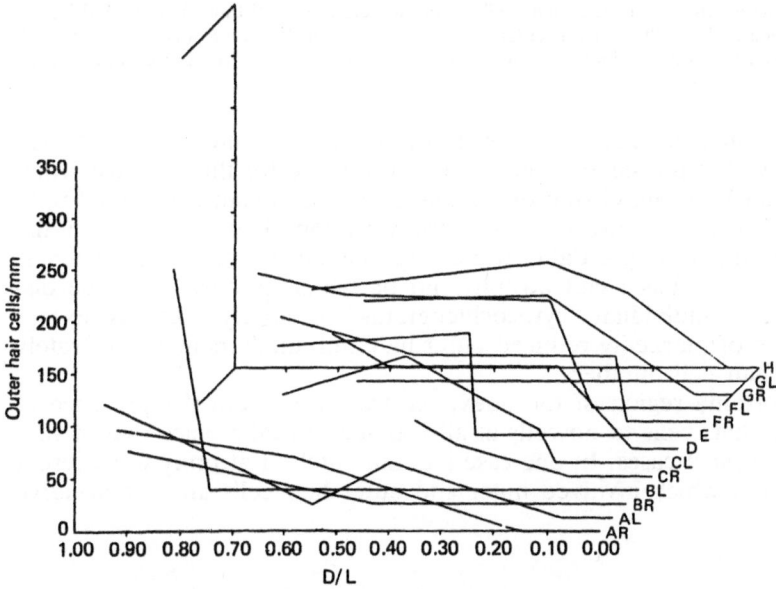

Fig. 4.14. Perspective curves of inner hair cell counts in relation to distance/length (D/L) from basal end of cochlea in 12 cochleas of elderly patients. *A,B,C*, etc. refer to individual patients. *R* and *L* refer to right and left cochleas, respectively, in those cases in whom both sides were examined. The curve of case GR is not shown, as inner hair cells could not be counted. To determine cell counts, points on the curves are referred to corresponding points on parallels to the vertical axis. *Initial horizontal lines* on the right indicate the areas of complete atrophy (Soucek et al. 1987).

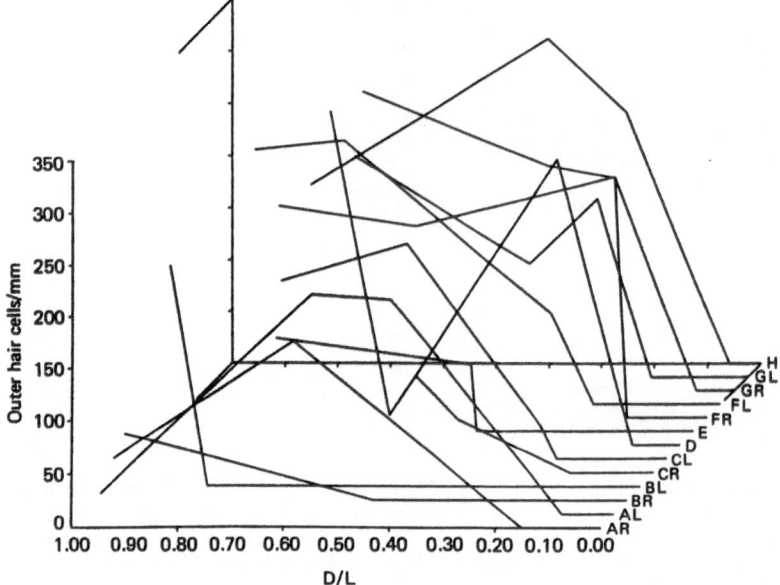

Fig. 4.15. Perspective curves of outer hair cell counts in relation to distance/length (D/L) from basal end of cochlea in 13 cochleas from elderly patients, drawn in the same way as in Fig. 4.14. The *initial horizontal lines* on the right indicate the areas of complete atrophy (Soucek et al. 1987).

single line for each cochlea. The levels of inner and outer hair cells are approximately within the same range as those produced for the 50+ age group by the European Working Group on normality of the human cochlea (Wright et al. 1987). There is, however, far more variation than is shown in the European Working Group's data, which are the means of many readings. Above the region of basal coil atrophy, no particular pattern could be discerned in these 13 individual "cytocochleograms" in elderly cochleas, except for the presence of markedly reduced outer hair cell numbers and moderately reduced inner hair cell numbers.

In the helicotrema region of the apex, the transition from the presence of inner and outer hair cells to no cells at all was seen to take place over a very short distance in most cases. In one case a 2 mm patch of atrophy was present at the apical end, which involved inner and outer hair cells and radial nerve fibres.

Discussion

The microslicing procedure was found to be very satisfactory for quantitative studies on surface preparations. The cochlea is "opened up" easily and quickly and there is little loss of cochlear tissue thanks to the precision of the cutting process. The vertical microslicing procedure, with construction of a diagram of the enlarged cochlear outline from the radiograph, allows a representation of the whole sequence of hair cell and nerve fibre appearances, throughout the length of the cochlea, to be charted.

The estimations of outer hair cells carried out in this investigation confirm the non-numerical interpretation given above that there is a marked reduction in numbers of these cells in the elderly. Third row cells suffer greater losses than those of the first row, and the basal coil is the most severely damaged. The inner hair cells are more susceptible to post-mortem autolysis, but also seem to suffer loss as part of the process of ageing, with giant stereociliary degeneration.

The most serious feature of the hearing loss which afflicts all elderly subjects is the deterioration of perception in the higher frequency range. The morphological counterpart of this appears to be complete atrophy of the terminal part the basal coil found in all elderly cochleas. The variable and usually moderate depression of hearing at other frequencies can be accounted for by the severe outer and moderate inner hair cell losses. Radial nerve fibre atrophy is a late stage which seems to follow hair cell disappearance. It is possible also that diminution of spiral ganglion cells which has been frequently described in presbyacusis is similarly related to longstanding total hair cell atrophy.

Summary of Histopathological Changes

This study was made on the hair cells of 57 cochleas from 39 patients over 65 years of age. As controls nine cochleas from six children less than 1 year of age were examined. In each cochlea perilymphatic perfusion–fixation was carried out, usually within 10 h of death. Surface preparations were made by sampling of microsliced preparations and stained by an osmic acid–Alcian blue eosin–phloxine method for ordinary light microscopy. Counts and measurements were made in some; others were assessed by approximation of the degree of change.

All cochleas showed hair cell changes of marked degree. There was severe, but not complete, loss of outer hair cells throughout all three cochlear coils. This was most marked in the third row of outer hair cells and least marked in the first row. There was mild loss of inner hair cells in the elderly throughout all cochlear coils.

In all cochleas a complete loss of both outer and inner hair cells was present at the extreme basal end. This varied in length from 2 to 9 mm. It was always acompanied by loss of radial nerve fibres.

Giant stereociliary degeneration, a marked elongation and fusion of stereocilia, was found to be present to a severe degree in apical and middle coil outer hair cells, and to a mild degree in inner hair cells of all three coils. Outer hair cells of basal coil did not show this change.

Fifteen of the 39 elderly patients had had pure tone audiometry and five auditory brainstem responses (ABR). Changes present were those characteristic of all elderly patients and described in Sections 2 and 3. It is suggested that the high tone loss is related to the severe terminal basal coil atrophy. The moderate loss throughout all frequencies and various ABR changes, and also the electrocochleographic changes described in Section 3, are related to severe outer hair cell damage.

In 15 of the elderly cochleas examined by surface preparation, paraffin sections of modiolus showed mild spiral ganglion cell losses only which could not be correlated with degree of basal coil atrophy. Light microscopic changes in sections of the stria vascularis could not be identified.

It is concluded that hearing loss in the elderly is essentially a hair cell lesion, which may be the culmination of many years of ageing degenerative changes.

References

Bredberg G (1965) Cellular pattern and nerve supply of the human organ of Corti. Acta Otolaryngol (Suppl) 236: 1–135

Brownell WE, Bader CR, Bertrand D, Ribaupierre Y de (1985) Evoked mechanical responses of isolated cochlear outer hair cells. Science 227: 194–196

Engström B, Flock A, Borg E (1983) Ultrastructural studies of stereocilia in noise-exposed rabbits. Hear Res 12: 251–264

Ernstson S (1971) Cochlear morphology in a strain of the waltzing guinea pig. Acta Otolaryngol (Stockh) 71: 469–482

Iurato S, Bredberg G, Bock G (1982) Functional histopathology of the human audio-vestibular organ. Eurodata hearing project. Commission of the European Communities

Johnsson LG, Hawkins JE (1967) A direct approach to cochlear anatomy and pathology in man. Arch Otolaryngol 85: 599–613

Johnsson LG, Hawkins JE (1972) Sensory and neural degeneration with ageing, as seen in microdissections of the human inner ear. Ann Otol Rhinol Laryngol 81: 179–193

Kirkwood TBL (1984) Towards a unified theory of cellular ageing. Monogr Dev Biol 17: 9–20

Michaels L (1987) Ear, nose and throat histopathology. Springer, London

Michaels L, Wells M, Frohlich A (1983) A new technique for the study of temporal bone pathology. Clin Otolaryngol 8: 77–85

Neely ST (1985) Micromechanics of the cochlear partition. In: Levin S, Allen JB, Hall JL, Hubbard A, Neely SF, Tubis A (eds) Peripheral auditory mechanisms proceedings, Boston. Springer, Berlin (Lecture notes in biomathematics, vol 64)

Nomura V, Kawabata I (1978) The pathology of sensory hairs in the human organ of Corti. Scanning Electron Microscopy 11: 417–422

Nomura Y, Kawabata I (1979) The loss of stereocilia in the human organ of Corti. Arch Otorhinolaryngol 222: 181–185

Saunders JC, Schneider ME, Dear SP (1985) The structure and function of actin in hair cells. J Acoust Soc Am 78: 299–311

Soucek S, Michaels L, Frohlich A (1986) Evidence for hair cell degeneration as the primary lesion in hearing loss of the elderly. J Otolaryngol 15: 175–183

Soucek S, Michaels L, Frohlich A (1987) Pathological changes in the organ of Corti in presbyacusis as revealed by microslicing and staining. Acta Otolaryngol (Stockh) (Suppl): 93–101

Spoendlin H, Baumgartner H (1977) Electrocochleography and cochlear pathology. Acta Otolaryngol 83: 130–135

Tilney LG, Saunders J (1983) Actin filaments, stereocilia and hair cells of the bird cochlea. Length, number, width and distribution of stereocilia of each hair cell are related to the position of the hair on the cochlea. J Cell Biol 96: 807–821

Wright A (1982) Giant cilia in the human organ of Corti. Clin Otolaryngol 7: 193–199

Wright A, Davis A, Bredberg G, Úlehlová L, Spencer H (1987) Hair cell distribution in the normal human cochlea. A report of a European Working Group. Acta Otolaryngol (Stockh) (Suppl) 436: 15–24

Section 5

Summing-up

Two questions were posed when commencing this work:

1. Where does the disturbance giving rise to old age deafness reside?
2. What is its cause?

The literature review of Section 1 gives little support for a central origin of presbyacusis. The central nervous system abnormalities which are undoubtedly present in many of the elderly hard-of-hearing, are there, it is becoming increasingly realized, as concomitant, not causal, processes. Most literature sources now accept that the site of the disturbance is in the cochlea. Strong evidence from a few histopathological studies has long incriminated outer hair cell degeneration as the primary lesion of old age hearing loss. Nevertheless, there is widespread support for the concept that there are four different sites of the degeneration in the cochlea: spiral ganglion cells and nerves, stria vascularis, basilar membrane as well as hair cells. Cell loss at each of these sites is purported to give rise to a different form of presbyacusis. This concept has been based on some histopathological, but no physiological, evidence.

In Section 1 we note the favour with which an exogenous cause of old age hearing loss is currently regarded in the literature. Noise, cardiovascular disease, hyperlipoproteinaemia, blood hyperviscosity and even ototoxic drugs are among the explanations brought forward, but without convincing evidence. The audiometric studies that we have undertaken and which are described in Section 2 show that hearing loss is universally present in the aged of both sexes, that the degree of deafness increases with advancing age and that noise exposure seems to play no significant role in its causation. These features suggest that presbyacusis is an innate process "caused" by old age like the other manifestations of senility and not one produced by the common diseases of later life. Further evidence for this view is given in Section 2 in that groups of old people with medical backgrounds ranging from the ambulant to the seriously ill nevertheless show similar degrees of hearing loss. Indeed the ill elderly appear to have better hearing than the healthy.

The application of electrophysiology to old age hearing loss has, as described in Section 3, led us to a more definite localization of the functional abnormality that is prevalent in this disorder. The pattern of brainstem evoked responses has allowed us to exclude any cerebral disturbance. Features of a form of recruitment in brainstem evoked responses and electrocochleograms suggest localization in the cochlea. We have been able to show that action potentials, while weaker than in the young, are still ample in all the elderly. From this we infer that "neural" presbyacusis does not exist as a functional entity. Adaptation changes follow similar lines in the elderly to those exhibited by the young suggesting again that in the former the functional integrity of the nerve endings of the eighth cranial nerve at the hair cells is retained. Localization of the disturbance to hair cells and not to other cochlear sites in old age deafness is emphasized by the weakness of the cochlear microphonics in the aged, and their deterioration still further with advancing age betokens an increasing process. Both in brainstem evoked responses and in electrocochleograms there is a ubiquity of the old age alterations and a constancy in their type which supports the concept of a degenerative lesion in old age rather than one produced by extraneous disease.

Finally in a large series of cases we have examined the inner ear microscopically after it has been fixed in situ within a short time after death. Our histopathological studies, described in Section 4, indicate the universality of hair cell degeneration in the ageing cochlea. Other degenerative changes are found, but not in every case. The hair cell changes, comprising marked general outer hair cell loss and severe terminal basal coil hair cell loss explain well the audiometric and electrophysiological alterations. In giant stereociliary degeneration we have described what could be a stage in the degenerative process of presbyacusis, at least in part of the hair cell system of the organ of Corti, which might yield further information on the pathogenesis of old age deafness if studied by modern methods of molecular biology.

We may conclude that old age deafness has its primary site in the hair cells of the cochlea and that it is caused by a process of senescence in that region.

Subject Index